4th Edition
CRASH COURSE

SERIES EDITOR
Dan Horton-Szar
BSc(Hons), MBBS(Hons), MRCGP
Northgate Medical Practice, Canterbury, Kent, UK

FACULTY ADVISOR
Paul Simons PhD
Senior Lecturer in Inflammation, Division of Medicine
University College London, London, UK

Cell Biology and Genetics

Matt Stubbs BSc
Foundation Doctor, North Central Thames Foundation School,
London, UK

Narin Suleyman BSc
Foundation Doctor, North Central Thames Foundation School,
London, UK

MOSBY

ELSEVIER

Edinburgh London New York Oxford Philadelphia St Louis Sydney Toronto 2013

MOSBY
ELSEVIER

Commissioning Editor: Jeremy Bowes
Development Editor: Ewan Halley
Project Manager: Andrew Riley
Designer/Design Direction: Stewart Larking
Illustration Manager: Jennifer Rose
Icon Illustrations: Geo Parkin

First edition 1998

Second edition 2002

Third edition 2008

Fourth edition 2013

ISBN 978-0-7234-3622-5

British Library Cataloguing in Publication Data
A catalogue record for this book is available from the British Library

Library of Congress Cataloging in Publication Data
A catalog record for this book is available from the Library of Congress

Printed in China

Cell Biology and Genetics

First edition authors:

Emma Jones

Anna Morris

Second edition author:

Ania L Manson

Third edition author:

Joanne Evans

Contents

Series editor foreword

The Crash Course series was first published in 1997 and now, 15 years on, we are still going strong. Medicine never stands still, and the work of keeping this series relevant for today's students is an ongoing process. These fourth editions build on the success of the previous titles and incorporate new and revised material, to keep the series up-to-date with current guidelines for best practice, and recent developments in medical research and pharmacology.

We always listen to feedback from our readers, through focus groups and student reviews of the Crash Course titles. For the fourth editions we have completely re-written our self-assessment material to keep up with today's 'single-best answer' and 'extended matching question' formats. The artwork and layout of the titles has also been largely re-worked to make it easier on the eye during long sessions of revision.

Despite fully revising the books with each edition, we hold fast to the principles on which we first developed the series. Crash Course will always bring you all the information you need to revise in compact, manageable volumes that integrate basic medical science and clinical practice. The books still maintain the balance between clarity and conciseness, and provide sufficient depth for those aiming at distinction. The authors are medical students and junior doctors who have recent experience of the exams you are now facing, and the accuracy of the material is checked by a team of faculty advisors from across the UK.

I wish you all the best for your future careers!

Dr Dan Horton-Szar

Authors

Cell biology and genetics is a subject both essential to the medical curriculum yet also feared by medical students. Love it or hate it, you need to understand these topics. They are essential to the core concepts of medical science whilst also being implicated in the latest developments in clinical research.

This fourth edition of *Cell Biology and Genetics* is not just a streamlined version of its predecessor. Developments in this field move especially fast and we have summarized some of the exciting new prospects to arise in the past few years. We have included the latest developments from the human genome project as well as spin-offs such as the thousand genome project. The latest molecular tools and laboratory techniques are explained so that you can appreciate where the latest treatments for genetic disease and screening technologies have come from. Technologies that will affect those with genetic diseases, such as therapeutic siRNA, zinc finger nucleases and DNA microarray technology, are some of the new and exciting aspects of this subject added to this edition. With a view to exams, we have also written questions that are most likely to appear as part of a current UK medical school curriculum so that you can test your recall and sharpen up your exam technique.

We have produced this text with the aim of making cell biology and genetics both as easy to understand and as easy to remember as possible. Whether you're using this text the night before an exam or to gain clarity on a difficult topic, we hope that by enabling your understanding this book may kick-start a passion for this fascinating subject.

Matt Stubbs and Narin Suleyman

Faculty advisor

These are exciting times in cell biology and genetics. Enormous progress has been made in very recent years, and the scene is set for further dramatic advances. This is perhaps most striking in human genomics and molecular genetics. As I write, ten years have elapsed since publication of the draft human genome sequence and we have moved to a position where identification of genes and mutations underlying genetic disease is commonplace. Gene identification has provided a starting point for understanding of the molecular mechanisms of many diseases, and is of enormous value for the management of genetic disease as well as for the identification of drug targets and in drug development. Now, information about normal genetic variation and increasingly powerful analytical methods mean that the genetic basis of complex traits is being elucidated. Such traits include diseases influenced by many genes (e.g. cardiovascular and Crohn diseases) as well as the range of normal phenotypes (e.g. body mass index and height). We are rapidly approaching a time when it will be possible to obtain the complete genome sequence of an individual for $1000 or less, and with it the prospect of widespread predictive genetic testing. In cell biology, new appreciation of the plasticity of cellular phenotypes, understanding of stem cell biology and methods of manipulating cell phenotypes raise hopes for regenerative medicine through stem cell therapy and tissue engineering, for example.

Modern cell biology and genetics have already had a great impact on clinical practice. In spite of the many technical and ethical challenges, it is inevitable that recent advances (and many yet to come) will quickly feed through to the clinic. Against this background, understanding of cell biology and genetics has never been more important for medical practitioners. With its emphasis on clarity, easy digestibility and conciseness, this book seeks to help you efficiently to gain a good grounding in the more established basic cell biology and genetics, as well as an appreciation of the latest developments and the directions of travel. The new edition has been updated to include key recent advances; in particular Chapter 7: Tools in Molecular Medicine has been extensively revised. As in previous editions, there is a strong emphasis on clinical relevance, including discussion of ethical issues. I hope that you will enjoy your reading of this book, that you will find it useful now and that it will help you to make sense of future developments as they happen.

Paul Simons

Acknowledgements

We would like to thank Dr Marsha Morgan for nominating us for the task of writing this, and for all the help she has given us both. We would also like to thank Dr Paul Simons for his excellent guidance and support throughout. Thanks also to the team at Elsevier for all the help and encouragement.

Figure Acknowledgements

Figure 1.1 adapted from Medical Microbiology, 3rd edn., by C. Mims et al., Mosby, 2004, Fig. 2.1

Figure 1.7 adapted from Medical Microbiology, 5th edn., by P.R. Murray, M.A. Pfaller and K.S. Rosenthal, Mosby, 2005, Fig. 6.9

Figures 2.1, 4.4, 4.6, 4.13–4.16, 4.27, 5.3, 6.2, 6.34 adapted from Human Histology, 2nd edn., by A. Stevens and J. Lowe, Mosby, 1997

Figure 2.3 electron micrographs reproduced courtesy of Dr Trevor Gray

Figures 3.10, 3.13, 5.13, 6.10 adapted from Medical Biochemistry, 1st edn., by J. Baynes and M. Dominiczak, Mosby, 1999

Figure 3.16 adapted from Physiology, 4th edn., by R. Berne et al., Mosby, 1998

Figures 4.2, 4.5, 4.9, 4.10, 6.4 adapted from Medical Cell Biology Made Memorable by R. Norman and D. Lodwick, Churchill Livingstone, 1999

Figures 4.3, 4.26, 6.3, 6.12 adapted with permission from Molecular Biology of the Cell, 3rd edn., by B. Alberts et al., Garland Publishing, 1994. Reproduced by permission of Routledge, Inc., part of the Taylor & Francis Group

Figure 4.25 adapted with permission from Molecular Cell Biology, 2nd edn., by Darnell, Lodish and Baltimore, Scientific American Books, 1990

Figures 6.16, 6.26, 7.5, 8.6, 8.7 adapted from Thompson & Thompson Genetics in Medicine by R.L. Nussbaum, R.R. McInnes and H.F. Willard, WB Saunders, 2001

Figures 6.17A, 8.20 adapted from Clinical Medicine, 6th edn., by P. Kumar and M. Clark, WB Saunders, 2005, Figs. 4.8, 6.40

Figure 6.21 adapted from Biochemistry, 3rd edn., by L. Stryer, W.H. Freeman and Company, 1988

Figure 7.3 reproduced courtesy of Dr Steve Howe

Figures 7.2, 7.15, 7.17, 7.18, 8.26 adapted from Emery's Elements of Medical Genetics, 11th edn., by R. Mueller and I. Young, Churchill Livingstone, 2001

Figure 7.10 reproduced courtesy of Linda E. Ritter

Acknowledgements

Figure 7.11 reproduced courtesy of Dr Paul Scriven, GSTT

Figure 7.19 reproduced courtesy of Dr Kathy Mann

Figures 8.30B, 8.32 adapted from Robbins & Cotran Pathologic Basis of Disease, 7th edn., by V. Kumar, A. Abbas and N. Fausto, WB Saunders, 2004, Fig. 5.26

Figures 8.14, 9.5 adapted from Emery's Elements of Medical Genetics, 12th edn., P. Turnpenny and S. Ellard, Churchill Livingstone, 2005, Fig. 21.4A

Figures 6.18, 8.35 adapted from Medical Genetics, 3rd edn., by L. Jorde, J. Carey and M. Bamshad, Mosby, 2003, Fig. 13.9

Dedication

Thank you to my family and friends for their constant encouragement, especially Mum, Baba, my Nenes and Dedes, Ediz (for doing whatever it is you do) and Ashley (for giving me pep talks). I'd also like to thank Matt for being a fantastic friend and co-author.

Narin Suleyman

To Mr Simon Hughes (Mill Hill School) whose excellent teaching inspired my interest in genetics. A huge thank you also to Mum, Dad, Al and Tia for all your love and support; and an apology to my great friend and co-author Narin, whose inbox may never recover. Finally, in memory of my inspirational grandfather; and with thanks to all of my ancestors for their genetic contribution.

Matt Stubbs

Cell biology and genetics of prokaryotes

1

Objectives

By the end of this chapter you should be able to:
- Explain why prokaryotic cells are smaller than eukaryotic cells.
- Describe the key features of a prokaryotic cell.
- Understand the differences between bacterial transformation, transduction and conjugation.
- Appreciate the main sites of action of antimicrobial agents.
- Recognize the main classes of virus based on nucleic acid composition.
- Understand key stages in the viral lifecycle.
- Understand the key targets of antiviral chemotherapy.

PROKARYOTIC CELL

Prokaryotes are the simplest unicellular organisms. It is generally accepted that all living organisms evolved from a common prokaryotic ancestor.

HINTS AND TIPS

The basic molecular machinery of life has been conserved in all species. The enzymes that perform common reactions, such as glycolysis, in bacterial and human cells show significant homology at both the DNA and protein level.

All microorganisms lacking a membrane-bound nucleus (i.e. the various types of bacteria) are classified in the prokaryote superkingdom. Classically prokaryotic cells (Fig. 1.1) show the following features.

- A single cytoplasmic compartment containing all the cellular components.
- Cell division usually by binary fission.

In order for the prokaryotic cell to survive, molecules required for energy and biosynthesis must diffuse into the cell, and waste products must diffuse out of the cell across the plasma membrane. The rate of diffusion is related to membrane surface area. When the diameter of a cell increases:

- the cell volume expands to the cube of the linear increase
- the surface area only expands to the square of the linear increase.

Thus, small cells have a larger surface area to volume ratio than large cells. If prokaryotes expand above a certain size, the rate of diffusion of nutrients across the plasma membrane will not be sufficient to sustain the increased needs of its larger cell volume. Hence, prokaryotic cells are small.

HINTS AND TIPS

Remember: **P**rokaryotes are **p**rimitive. **You** are a **eu**karyote.

Prokaryotic cell structure and organelles

Plasma membrane

The prokaryotic cell membrane is a fluid phospholipid bilayer. With the exception of the mycoplasmas, they lack sterols, instead containing sterol-like molecules called hopanoids. The main function of the plasma membrane is to act as a selectively permeable membrane.

Cell wall

All bacteria (except mycoplasmas) possess a complex cell wall. In the true bacteria, this is usually a peptidoglycan structure, a polymer of amino sugars. As bacteria concentrate dissolved nutrients through active transport, their cytoplasm is usually hypertonic. The cell wall acts to prevent osmotic lysis of the cell and is, therefore, a common antibiotic target (see p. 7) (Fig. 1.2).

HINTS AND TIPS

Mycoplasmas are degenerate bacteria that do not have cell walls. They are the smallest known free-living life forms. Several species are pathogenic in humans, including *M. pneumoniae*, which is an important cause of atypical pneumonia and other respiratory disorders.

Clinical Note

Bacterial cell walls can be described according to their staining ability, primarily being *Gram-positive* (retain the crystal violet dye during the Gram-stain procedure), *Gram-negative* (decolorize during the Gram-stain procedure) and *acid fast*. Gram-positive bacteria have thick (20–80 nm) peptidoglycan walls external to the cell membrane, while Gram-negative species have a protected thin (5–10 nm) peptidoglycan layer, overlaid by a lipopolysaccharide-rich outer membrane. When coupled with cell shape, such properties aid organism identification; for example Gram-negative diplococci in cerebrospinal fluid are typical of meningococcal meningitis.

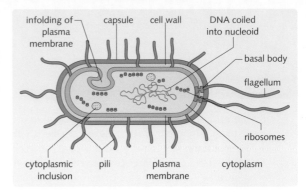

Fig. 1.1 Structure of a typical prokaryotic cell. The DNA molecule is free in the cytoplasm. Bacteria contain some sub-cellular structures, such as ribosomes, but do not contain membrane bound organelles.

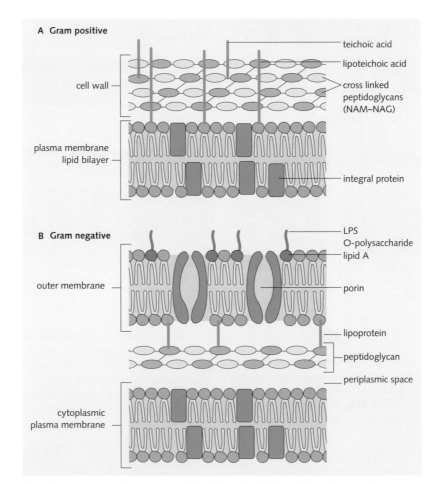

Fig. 1.2 Bacterial cell wall structure. (A) The Gram-positive cell wall is made mostly of peptidoglycan, interspersed with teichoic acid, which chelate the different layers together. (B) The Gram-negative cell wall contains a thin peptidoglycan layer with much less crosslinking between the peptidoglycan than seen in Gram-positive cell walls. The peptidoglycan layer is surrounded by an outer membrane, the outer leaflet of which contains antigenic lipopolysaccharide (LPS). In most cases this membrane structure is anchored non-covalently to lipoprotein molecules, which are covalently linked to the peptidoglycan.

Along with the cytoskeleton, the cell wall maintains the overall shape of the bacterial cell. The most common shapes held by bacteria are:

- coccus (spherical or oval)
- bacillus (rod-shaped)
- helical.

Ribosomes

Ribosomes are small cellular components composed of ribosomal RNA (rRNA) and ribosomal protein (ribonucleoprotein). They are the site of protein synthesis (translation). Bacterial ribosomes are composed of two subunits with densities of 50 S and 30 S. The two subunits combine during protein synthesis to form a complete 70 S ribosome, about 25 nm in diameter. The smaller size of the bacterial ribosome, compared to the 80 S eukaryotic ribosome, makes it an ideal antimicrobial target.

Nucleiod

The bacterial genome is encoded by a single chromosome also known as the nucleoid. A structure measuring approximately 0.2 μm in diameter, it is formed from supercoiled DNA and histone-like proteins. It is not membrane bound and, therefore, is free in the cytoplasm. Bacterial DNA lacks introns, instead comprising a continuous coding sequence of genes, usually clustered into functional units called operons (see p. 81).

Cytoskeleton

It was long assumed that prokaryotic cells did not contain a cytoskeleton; however, recently at least two major components of the bacterial cytoskeleton have been identified: the bacterial tubulin and actin homologues FtsZ and MreB. FtsZ is thought to be involved in cell division, while MreB proteins are thought to be involved in the regulation of cell shape and the segregation of some bacterial plasmids.

Cell specializations

Glycocalyx

The glycocalyx, forming the capsule or slime layer, is a mucopolysaccharide sheet external to the cell wall. Where present, the main functions of the glycocalyx are to protect against phagocytic engulfment by the host cell and to enable bacteria to adhere to and colonize surfaces.

Clinical Note

Streptococcus pneumoniae is a spherical, Gram-positive human pathogen. Infection can lead to pneumonia, meningitis, septic arthritis, endocarditis and otitis media, among others. The virulence of these organisms is associated with the presence of a polysaccharide capsule. Mutants unable to synthesize a capsule are not pathogenic to humans; loss of the capsule is accompanied by a 10^5-fold reduction in virulence. As asplenic patients are particularly at risk from infection with capsulated organisms such as pneumococcus, with an associated mortality of around 60%, they should be offered pneumococcal vaccination.

Flagella

Flagella are long structures extending from the cell surface, enabling bacteria to move in their environment. The bacterial flagellum is made of the protein flagellin and is a hollow tube 20 nm in diameter. It has three parts:

- basal body
- hook
- filament.

The basal body consists of a reversible rotary motor embedded in the cell wall, beginning within the cytoplasm and ending at the outer membrane. The hook is a flexible coupling or universal joint, and the long helical filament acts as a propeller (Fig. 1.3).

Pili

Pili are thin, protein tubes originating from the cytoplasmic membrane. They are found almost exclusively in Gram-negative bacteria. Pili are more rigid than flagella and have a role in bacterial attachment, either to another bacterium via 'sex' pili or to a host cell via the 'common' pili. The presence of many pili is thought

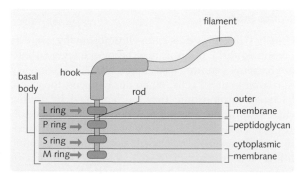

Fig. 1.3 Schematic of a flagellum of a Gram-negative bacteria. The basal body acts as a molecular motor, enabling the flagellum to be rotated, and consists of a rod and a series of rings that anchor the flagellum to the cell wall and the cytoplasmic membrane. While Gram-negative organisms have four basal rings, Gram-positive organisms only have two: one in the peptidoglycan layer and one in the plasma membrane. The hook provides a flexible coupling between the filament and the basal body. The filament is a hollow tube composed of the protein flagellin, ending with a capping protein.

to prevent phagocytosis, reducing host resistance to bacterial infection.

TRANSFER OF GENETIC MATERIAL

Although bacteria are not capable of true sexual reproduction, they may exchange genetic material by three mechanisms.

- Transformation – certain bacteria release DNA into the environment that can be taken up by other bacteria via specific receptors. If the DNA is compatible, it is incorporated into the bacterial genome; otherwise it is degraded by exonucleases.
- Transduction – a fragment of bacterial DNA may become incorporated into a bacteriophage (a virus that infects bacteria) during its assembly. One in 10^6 bacteriophages contain such bacterial DNA. This may be introduced into the host bacterial cell along with viral genes during the infection process. Again, if the bacterial DNA is compatible it can be integrated into the host's genome.
- Conjugation – this is sometimes loosely called bacterial sexual reproduction (Fig. 1.4). Bacteria may be designated F positive (F^+) or F negative (F^-). F^+ cells possess a plasmid called the F factor, which includes genes for a 'sex' pilus, giving F^+ cells the ability to attach to other bacterial cells. In this way it forms cytoplasmic bridges through which genetic material may then be transferred after it has been replicated by 'rolling circle replication'. Other plasmids present, for example those conferring antibiotic resistance, may also be transferred during this process.

Usually F plasmids exist extra-chromosomally. On rare occasions, the F plasmid may become integrated into the bacterial genome, resulting in Hfr (high frequency of recombination) cells. When this happens, the whole bacterial chromosome may be replicated by rolling circle replication, starting at an origin in the F factor. The replicated DNA, that now includes bacterial chromosomal material, is passed along the cytoplasmic bridge into the recipient cell. Variable amounts of DNA are transferred because the bridge invariably breaks down before the whole chromosome has been transmitted. The transmitted material may then undergo recombination with the host chromosome.

Infrequently, an F plasmid that has integrated into the genome may 'pop-out' again, taking part of the bacterial chromosome with it. These are called F′ plasmids and they may act as a vehicle to transmit bacterial genes to new hosts.

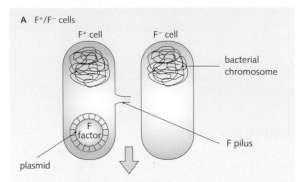

A F^+/F^- cells

F+ cell · F− cell · bacterial chromosome · F factor · plasmid · F pilus

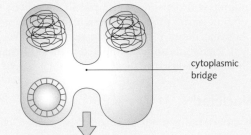

B cytoplasmic bridge formation via F pilus

cytoplasmic bridge

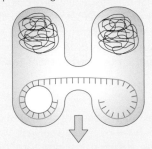

C F plasmid replication and transfer to F⁻ cell via cytoplasmic bridge

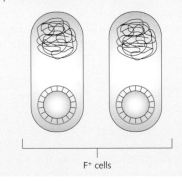

D plasmid transferred. Both cells now F+

F+ cells

Fig. 1.4 Bacterial conjugation. (A) The ability to conjugate is conferred by a plasmid called the F factor. Bacteria with the plasmid are F^+. (B) F^+ cells contain thread-like projections called F pili, which can attach to F-cells to form a cytoplasmic bridge. (C) F plasmid is replicated and a single-stranded replica is transferred along the bridge. (D) Within the recipient, the transferred material is replicated to form a new plasmid.

The conjugation process provides bacteria with a means of acquiring genes that, although beneficial to the organism, are not to their hosts. During conjugation, the F^+ cell can also pass an 'R plasmid', containing several antibiotic resistance genes, to the recipient F^- bacterium. The recipient bacterium is now not only antibiotic resistant, but also capable of producing a sex pilus (F^+) and passing on antibiotic resistance to surrounding F^- cells.

DNA REPLICATION

DNA replication is the process by which double-stranded DNA molecules are divided longitudinally, such that each strand is conserved to act as a template for the formation of a new strand. It is said to be semi-conservative, since only one strand is newly synthesized in each daughter molecule.

The bacterial nucleoid does not divide by mitosis. With its single chromosome and a division time of 20 min at 37°C, replication has been studied extensively in *Escherichia coli* (Fig. 1.5). It has been possible to isolate a range of replication deficient mutants, which have been used to identify and characterize the corresponding replication proteins. Such studies suggest that even prokaryotic replication is a complex process that requires about 30 proteins.

DNA polymerases

DNA polymerases are the enzymes responsible for DNA-chain synthesis. They couple nucleoside triphosphates onto a growing DNA strand by adding a phosphate group onto the free 3'-OH group. New DNA molecules are thus synthesized in a 5'-3' direction. Polymerization is driven thermodynamically by the elimination of a pyrophosphate (PP_i) and its subsequent hydrolysis:

$$(DNA)_n + dNTP \rightarrow (DNA)_{n+1} + PP_i$$

Three DNA polymerases have been characterized in *E. coli*, of which two are important in replication. RNA polymerase can begin a polynucleotide chain by linking two nucleoside triphosphates together directly (see p. 74). In contrast, DNA polymerase has an absolute requirement for a perfectly base-paired nucleotide, onto which it can then add nucleotides at the 3'-OH end. This has important consequences.

- DNA polymerase requires a primer on which to initiate extension.
- It will pause if an incorrect base is inserted.

The primers that are required for DNA polymerase activity are synthesized by RNA polymerase, since this enzyme does not itself require priming oligonucleotides. It synthesizes short stretches (10–20 nucleotides) of RNA primer sequence.

If Pol III, the main replication enzyme, inserts an incorrect base into the extending DNA chain, it cannot proceed until the erroneous nucleotide is excised.

Fig. 1.5 DNA polymerases in *Escherichia coli*.

	DNA Pol I	DNA Pol III
Notes	First polymerase to be discovered by Kornberg in 1957	Discovered when a mutant strain of *E. coli* with very low Pol I activity was shown to have a normal rate of reproduction
Structure	Single polypeptide with 928 residues, 109 kDa mass; forms one large ('Klenow') fragment and one small fragment	Three subunits with total 140 kDa; subunits—a, e, q
Functions		Polymerase (a subunit), 3'–5' exonuclease (e subunit) 5'–3' exonuclease
Associated proteins	Nil	At least seven other proteins associate to form a complex called the Pol III holoenzyme
Processivity (the number of consecutive reactions the enzyme is capable of performing)	At least 20 consecutive polymerization steps can occur before Pol I becomes dissociated from the DNA	In the holoenzyme the extra subunits interact with DNA and other proteins to clamp the polymerase onto the DNA, creating very high processivity (> 5000 residues); Pol III alone has a processivity of 10–15 residues
Main biological function	Proofreading and error correction	DNA replication, proofreading

This is achieved by its own 3′–5′ exonuclease activity. In this manner, the enzyme corrects its own mistakes as it goes along. This function is called 'proofreading', and it maintains the fidelity of DNA sequence after replication. Errors in replication occur at a rate of about 1 in 10^5 base pairs, reduced to about 1 in 10^8 by proofreading mechanisms.

DNA replication fork

Replication is initiated at the 'origin of replication', giving rise to two replication forks. Replication complexes assemble at both of these and proceed in opposite directions (Fig. 1.6).

Since DNA-dependent polymerases can only add nucleotides to the 3′ end, the leading strand can be synthesized in a continuous process from a single RNA primer. However, the lagging strand must be synthesized in pieces (called Okazaki fragments), which are subsequently joined together by a DNA ligase (Fig. 1.6).

Prokaryotic cell replication, transcription and translation involve phases of initiation, elongation and termination similar to that of eukaryotic cell replication. These stages are described, in the eukaryote, in Chapter 7. The important differences between prokaryotic and eukaryotic cells are listed in Figure 2.2 (see p. 14).

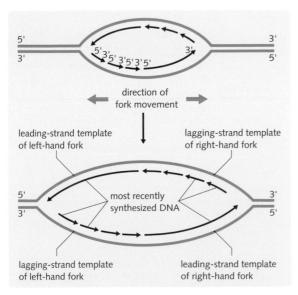

Fig. 1.6 The replication bubble. Two replication complexes form at each origin of replication and proceed in opposite directions. Note that the strand that is leading and the strand that is lagging depends upon the direction the replication complex is migrating.

ANTIMICROBIAL AGENTS

There are fundamental differences in the cellular machinery of bacterial and mammalian cells, and mammals can tolerate some chemicals that are toxic to bacteria. Therefore, humans can generally take antimicrobial agents in appropriate amounts to treat bacterial infections without harming themselves. A summary of the differences between bacterial and human cells is given in Figure 2.2 (see p. 14).

There are three ways of classifying antimicrobial agents.

1. By target site, which is the most practical.
2. By chemical structure.
3. According to whether they are bactericidal (kill) or bacteriostatic (inhibit growth).

Antimicrobial agents that kill bacteria are irreversible; those that inhibit growth are reversible. Some agents may act in different ways depending on the organism being treated, so what may be bactericidal in one organism, may only be bacteriostatic in another.

> **Clinical Note**
>
> Bacteriostatic agents include sulphonamides, tetracyclines, chloramphenicol and erythromycin. In order to be effective, they require a competent immune system. The bacteriostatic agent prevents the bacterial population from increasing, allowing the host immune system to eliminate the remaining pathogens. Therefore, the duration of therapy must be sufficient to allow cellular and humoral defence mechanisms to become active. Unsurprisingly, in the case of immunocompromised patients, antibiotics with bactericidal activity are, therefore, preferred.

There are five main target sites for antibacterial agents.

1. Nucleic acid synthesis.
2. Cell wall synthesis.
3. Protein synthesis.
4. Metabolic pathways.
5. Cell membrane function.

Inhibitors of nucleic acid synthesis

Significant differences between eukaryotic and prokaryotic replication allow for the differential targeting of prokaryotic replication. Nucleic acid synthesis can be inhibited at the level of:

- DNA replication
- RNA polymerase activity
- inhibition of nucleic acid precursors (see p. 74).

DNA replication

Quinolones (e.g. ciprofloxacin) inhibit bacterial DNA gyrase and topoisomerase IV, which are the enzymes responsible for unwinding and rewinding supercoiled DNA either side of the replication fork, thus inhibiting DNA replication. They are bactericidal and are selective for bacteria, as they do not affect the mammalian versions of these enzymes.

RNA polymerase activity

Rifampicin inhibits RNA synthesis by inhibiting DNA-dependent RNA polymerase, blocking the synthesis of mRNA. All members of the rifampicin family are bactericidal and show selective toxicity with a greater affinity for bacterial polymerases than for the equivalent human enzymes.

Inhibitors of cell-wall synthesis

The bacterial cell wall is rigid, containing linear peptidoglycans that are cross-linked by peptides (see Fig. 1.2). The absence of such molecules in mammalian cell membranes allows them to be targeted. Disrupting the cell wall leaves the bacteria susceptible to osmotic lysis and death.

β-lactams

Penicillins were the first group of antibiotics to be discovered and they are still important clinically. Penicillin itself is active mainly, but not exclusively, against Gram-positive organisms, whereas synthetic penicillins have been developed for their activity against Gram-negative rods.

Benzylpenicillin:

- is used in the treatment of pneumococcal, other streptococcal and meningococcal infections. However, there have been reports suggesting the emergence of benzylpenicillin-resistant pneumococci
- is not effective orally, so it is given by injection.

Cephalosporins are also β-lactams and have the same mode of action as penicillin. First-generation cephalosporins were most active against Gram-positive bacteria, but modified second- and third-generation drugs have a much broader spectrum of activity, which includes Gram-negative bacteria. They are also resistant to β-lactamase. Thus, cefuroxime (second-generation) is active against *S. aureus*, and ceftazidime (third-generation) has activity against the *Pseudomonas* species that can cause infection in immunocompromised individuals.

Glycopeptides

The glycopeptides vancomycin and teicoplanin inhibit peptidoglycan synthesis by acting at an earlier stage than β-lactams. They covalently bind to terminal D-alanine–D-alanine at the end of pentapeptide chains, thus sterically inhibiting the elongation of the peptidoglycan backbone by preventing the incorporation of new subunits into the growing cell wall. They are only effective against Gram-positive organisms, as their large size means they cannot penetrate easily into Gram-negative cells. As they are expensive and potentially toxic, glycopeptides are reserved for severe infections, for infections with organisms that are resistant to other antibiotics, or in cases where a patient has displayed hypersensitivity to β-lactams.

Inhibitors of protein synthesis

Many antibiotics block protein synthesis, either by blocking translation or by other means. Subtle difference between prokaryotes and eukaryotes, for example in ribosome size, mean that selectivity is possible. For example, protein synthesis inhibitors, such as tetracycline, kanamycin and erythromycin, target prokaryotic ribosomes, but they do not affect mammalian ribosomes. Inhibition can be effected at all stages of translation from initiation to elongation to termination.

> **HINTS AND TIPS**
>
> To recall antibiotics that interact with the different subunits of the bacterial ribosome remember 'Buy *AT* 30, *CELL* at 50'. A=*A*minoglycosides; T=*T*etracycline; C=*C*hloramphenicol; E=*E*rythromycin; L=*L*inezolid; L=c*L*indamycin.

Antibiotics and mitochondria

Prokaryotic cells do not possess membrane-bound organelles and, as such, do not possess mitochondria. However, it is generally accepted that eukaryotic mitochondria originated as bacterial endosymbionts, evolving from prokaryotic cells and, as such, the translation process in eukaryotic mitochondria is very similar to that in prokaryotic cells. Antibiotics that inhibit prokaryotic protein synthesis can also affect mitochondrial protein synthesis. Antibiotics, however, do not harm their mammalian host because:

- some antibiotics are unable to cross the inner mitochondrial membrane
- mitochondria are replaced at cell division. This occurs relatively slowly in most cells, so mitochondria are depleted only with long-term antibiotic use
- in rapidly dividing cells, the local environment can sometimes prevent uptake of antibiotic (e.g. in bone, high calcium levels cause the formation of calcium-tetracycline, so the drug cannot be taken up by cells in the bone marrow).

Inhibitors of metabolic pathways: anti-metabolites

These agents:

- target the folic acid synthesis pathway (this pathway produces tetrahydrofolate, which is essential for nucleotide synthesis)
- do not affect mammalian cells because mammals obtain folic acid from their diet
- are bacteriostatic.

Sulphonamides (e.g. sulphadiazine) are analogues of γ-aminobenzoic acid. Trimethoprim inhibits bacterial, but not eukaryotic, dihydrofolate reductase. It is used in the treatment of urinary-tract infections.

Inhibitors of cell-membrane function

The cell membrane controls the internal composition of the cell, and the disruption of the cell membrane can lead to changes in membrane function and permeability, leading to cell damage or death. The polymyxins are active against all Gram-negative organisms, except *Proteus* species. They act as cationic detergents, disrupting the phospholipid structure of the cell membrane. They can only be given systemically and, because they are toxic, have few indications.

Antibiotic resistance

Antibiotic resistance in bacteria may be a natural characteristic of the organism (e.g. it may lack the target of the antibiotic molecule) or may be acquired. Acquired resistance is a major problem clinically and is brought about by mechanisms that can be broadly classified into three main types.

- Alteration of the antibiotic – some bacteria can produce an enzyme that can chemically alter the structure of the antibiotic. As the structure of the antibiotic is essential to its interaction with its target molecule, this neutralizes the antibiotic before it can have an effect. Examples include the β-lactamases, aminoglycoside-modifying enzymes and chloramphenicol acetyl transferases.
- Target-mediated resistance – the target, for example a receptor, may be altered so that it has a lowered affinity for the antibacterial agent in question.
- Impermeability – bacteria can reduce the amount of drug that reaches the target either by decreasing the permeability of the cell wall to the antibiotic or by pumping the drug out of the cell (known as an efflux mechanism).

Bacteria can acquire resistance by a number of mechanisms, including chromosomal mutation leading to class resistance; the horizontal transfer of resistance genes by conjugation, transformation, transduction; or by the acquisition of 'jumping genes' (transposable elements) or 'cassettes' of resistance genes (integrons).

The use of two or more antibiotics, known as combination therapy, can be used in special cases of antibiotic resistance. This approach is useful in fighting the infection as antibiotics can work synergistically and provides broad coverage in emergency infections of unknown aetiology. The use of combination therapy also prevents resistant strains developing.

VIRUSES

Viruses are infectious particles consisting of nuclear material enclosed in a protein coat called a capsid, which may be surrounded by a phospholipid envelope. The complete virus particle is known as a virion. Viruses are obligate intracellular parasites, being totally dependent on the cells that they infect to provide metabolic intermediates, energy and many, if not all, of the enzymes they require to replicate. Generally they infect their host by means of receptor-mediated endocytosis or fusion with the plasma membrane. The major stages in viral replication are generally the same for all viruses (Fig. 1.7).

HINTS AND TIPS

One of the most significant challenges in gene therapy is getting the therapeutic piece of DNA through the plasma membrane to the nucleus, while avoiding lysosomal degradation. Viruses have evolved very efficient mechanisms for doing this and, therefore, many gene therapy protocols exploit viral vectors.

Replication of RNA viral genomes is error-prone and leads to genome diversity. This is especially relevant for the development of antiviral resistance by HIV. In contrast, replication of DNA genomes is relatively error-free due to the proofreading activity of viral DNA polymerase.

Viral genomes

Viral genomes vary greatly in size, from approximately 3200 nucleotides (hepadnavirus) to approximately 1.2 million base pairs (mimivirus). They contain either deoxyribonucleic acid or ribonucleic acid, but not

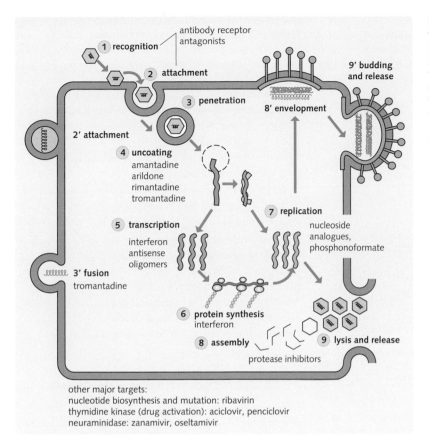

Fig. 1.7 Lifecycle of a virus. In order to replicate, each virus must infect a host cell and 'hijack' its cellular machinery. For the cycle to continue, the newly assembled virions must escape from the original host cell so that they can go on to infect new host cells. The ten stages that the virus must successfully pass through to complete this cycle are shown.

usually both. The exception to this is cytomegalovirus (CMV), which contains both a DNA core and mRNA.

DNA viruses

Double-stranded DNA viruses
The genome is a molecule of double-stranded (ds) DNA that consists of a 'plus' strand and a 'minus' strand. The 'minus' DNA strand is directly transcribed into viral mRNA. Important examples include the herpes family of viruses and adenoviruses.

Single-stranded DNA viruses
The genome is a molecule of single-stranded (ss) DNA. Once inside the host cell, the ssDNA is converted into dsDNA, and the minus DNA strand is transcribed into viral mRNA. Examples of ssDNA virus include the adeno-associated virus and parvovirus B19.

Reverse transcribing DNA viruses
Although the genome is double-stranded DNA, its replication involves the generation of an intermediate mRNA molecule, known as the pre-genome, from which DNA is reverse transcribed. Hepatitis B can be found in this group.

RNA viruses

Double-stranded RNA viruses
The genome is a molecule of double-stranded RNA. The 'plus' RNA strand functions as viral mRNA. An example is the reovirus family, which includes rotavirus.

Positive-sense, single-stranded RNA viruses
Positive-sense viral RNA is identical to viral mRNA, and as such can be immediately translated into viral proteins. Examples include polioviruses, rubella and the hepatitis A and C viruses.

Negative-sense, single-stranded RNA viruses
Negative-sense viral RNA is a mirror image of mRNA and must be converted to positivesense RNA by a virally encoded RNA-dependent polymerase before viral proteins can be translated. Examples include the influenza, measles, mumps and Ebola viruses.

Reverse transcribing RNA viruses

Also known as retroviruses, these are positive-sense ssRNA viruses that produce a DNA inter mediate with a unique enzyme called reverse transcriptase (Figs 1.8 and 1.9).

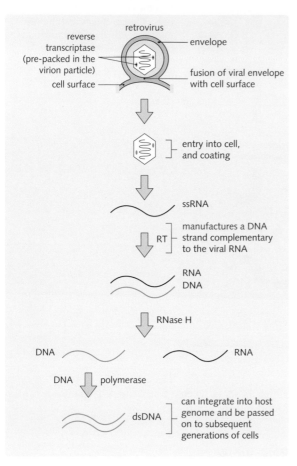

Fig. 1.8 Replication of retroviruses.

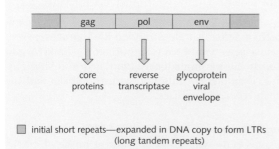

Fig. 1.9 Prototype retrovirus genome – codes for three proteins. The genes are preceded by a non-coding region containing enhancer and promoter regions that facilitate expression of the viral genome by host 'machinery'.

Examples include the human immunodeficiency viruses (HIV) and human T lymphocyte viruses 1 and 2.

> **Clinical Note**
>
> Human immunodeficiency viruses HIV-1 and HIV-2 are retroviruses and are responsible for the worldwide pandemic of acquired immune deficiency syndrome (AIDS). HIV is transmitted:
> - in blood (needlestick injury, sharing needles, contaminated blood products)
> - by sexual contact (heterosexual and homosexual, and from unscreened sperm donors)
> - from mother to baby (during pregnancy, childbirth or breastfeeding).
>
> HIV attacks the immune system and progressively destroys it. It does this by specifically targeting $CD4^+$ cells (T helper cells, dendritic cells and macrophages) by binding onto the CD4 receptor. The viral DNA integrates into the host genome and new virus particles are manufactured and mature HIV virions released.

Pathogenesis of viral infection

Following infection of a cell, virus assembly, maturation and release lead to a wide range of virus–cell interactions, which can be grouped into three main categories.

1. Lytic – following infection, the host cell is lysed in order to release the new viral progeny.
2. Persistent – the virus infects the host cell, but does not complete the replication cycle.
3. Lysogenic – viral DNA becomes integrated with that of the host and, as such, can be transmitted to daughter cells at each subsequent cell division. This can result in latency or transformation.

Although useful for classifying infection, these categories are not mutually exclusive, and viruses can be capable of more than one lifecycle. For example, lytic infections by human papillomavirus (HPV) cause genital warts, while latent infections by some strains of HPV lead to cervical and anogenital cancer.

Antiviral chemotherapy

Antiviral drugs work by interfering with viral replication. Since viruses utilize the host cells' own metabolic pathways for replication, it is difficult to find drugs that are virus-specific. Nevertheless, there are enzymes that are only encoded by the virus, which offer potential virus-specific targets. As such, most antiviral agents are only effective while the virus is replicating (with reference to numbered stages on Fig. 1.7):

- viral attachment to the host cell (2)
- viral penetration (3) and uncoating (4)
- viral nucleic acid synthesis (6)
- virus particle maturation (9)
- virus release (9).

Viral attachment to the host cell

By interfering with the binding and adsorption of viral particles into the host cell (see Fig. 1.7), the host cell is protected from viral infection. Although largely an experimental target, one fusion inhibitor, T-20, has been approved in the US for the treatment of HIV. T-20 prevents viral attachment to the T cell by binding to the HIV transmembrane glycoprotein site, gp41.

Viral penetration and uncoating

After attachment, the virus penetrates the host cell within an endocytic vesicle. This is followed by 'uncoating', during which the endocytic vesicle and the viral capsid are enzymatically degraded and the viral genome is released (see Fig. 1.7). By inhibiting this stage, the viral genome is not released and cannot take over the host's translational machinery. Amantadine prevents the influenza A virion uncoating by interacting with the viral membrane ion channel, M2, sterically blocking it.

Viral nucleic-acid synthesis

Viruses encode specific enzymes of their own. For example, polymerases are often virally encoded. These include:

- DNA-dependent DNA polymerase – DNA viruses
- RNA-dependent RNA polymerase – RNA viruses
- RNA-dependent DNA polymerase – retroviruses (Figs 1.8, 1.9).

Other targets include enzymes required for nucleic acid synthesis, for example herpes simplex thymidine kinase.

DNA polymerase inhibitors

Aciclovir is an acyclic nucleoside analogue that acts as a chain terminator of herpesvirus DNA synthesis. It is a prodrug, converted to aciclovir monophosphate by α-herpesvirus encoded thymidine kinase. Subsequent conversion to aciclovir triphosphate is achieved by host cellular enzymes. Aciclovir triphosphate then competes with deoxyguanosine triphosphate for incorporation into viral DNA, resulting in DNA synthesis termination.

Foscarnet is a pyrophosphate analogue that prevents nucleotide binding, thus inhibiting viral replication. It is active against herpesvirus and its main application is in the treatment of severe cytomegalovirus disease and for the treatment of aciclovir-resistant herpes simplex infection.

Reverse transcriptase inhibitors

Reverse transcriptase inhibitors (RTIs) have revolutionized the treatment of HIV. The main classes of RTIs are:

- nucleoside analogue reverse transcriptase inhibitors (NRTIs)
- nucleotide analogue reverse transcriptase inhibitors (NtRTIs)
- non-nucleoside reverse transcriptase inhibitors (NNRTIs).

NRTIs inhibit viral RNA-dependent DNA polymerase (reverse transcriptase) and are incorporated into viral DNA leading to chain termination. They require conversion into their active 'triphosphate' form to become active. Examples include zidovudine (also called azidothymidine, AZT), an analogue of thymidine, and didanosine (ddI), an analogue of deoxyadenosine, which are phosphorylated in the host cell and compete with cellular nucleotide triphosphates in the reverse transcription process. Host mitochondrial, γ DNA polymerase is also susceptible to NRTIs and this probably accounts for some of the common side effects. These include anaemia and neutropenia, gastrointestinal disturbance, alterations of liver function, headache, fever and skin rash.

NtRTIs differ from the NRTIs in that they require only one phosphorylation step within the cell to become activated, thus resulting in less toxicity. An example is tenofovir, the only NtRTI currently licensed for HIV treatment.

The NNRTIs have a different mode of action. They are not incorporated into viral DNA, but inhibit replication directly by binding non-competitively to reverse transcriptase. Examples include nevirapine and efavirenz.

Virus particle maturation

In order to be able to infect a new host cell, newly synthesized virions need to mature (see Fig. 1.7). The HIV protease is an essential enzyme in the maturation of the HIV virus. Protease inhibitors, such as saquinavir, nelfinavir and ritonavir, prevent viral assembly by inhibiting the activity of viral protease, the enzyme required to cleave HIV polyproteins into functionally active proteins. As a result, HIV cannot mature and noninfectious viruses are produced.

Virus release

The final stage in the lifecycle of a virus is the release of completed viruses from the host cell (see Fig. 1.7). Neuraminidase, one of two influenza surface glycoproteins, functions to cleave sialic acid residues from the host cell, releasing new virions. The neuraminidase inhibitors zanamivir and oseltamivir, which are active against both influenza A and B viruses, are competitive reversible inhibitors of the neuraminidase active site.

Eukaryotic organelles 2

● **Objectives**

By the end of this chapter you should be able to:
- Draw and label a typical prokaryotic and a typical eukaryotic cell.
- Explain why eukaryotic cells are larger than prokaryotic cells.
- Outline the differences between prokaryotes and eukaryotes.
- Describe epithelial cell membrane specializations.
- Understand how the structure of membranous organelles relates to function.
- List the three components of the cytoskeleton.
- Outline the structure and function of cilia.
- Describe two types of specialized cell and explain how the structure of each reflects its function.
- Explain cell differentiation and cell memory.

THE EUKARYOTIC CELL

All organisms consisting of cells with a membrane-bound nucleus are classified in the eukaryote super-kingdom. The animalia, plantae, protista, and fungi kingdoms all belong within this group. The typical eukaryotic cell (Fig. 2.1) shows the following features:

- a complex series of inner membranes that separate the cell into distinct compartments that perform specific functions
- cell division by mitosis
- specialized organelles, such as centrioles, mitotic spindles, mitochondria and microtubules.

Unlike prokaryotes, eukaryotic cells are capable of endocytosis (see Ch. 4). This allows patches of the plasma membrane to pinch off to form membrane-bound vesicles, delivering nutrients from the external environment to compartments deep within the cell. Endocytosis liberates eukaryotic cells from the constraints of simple diffusion, allowing them to sustain a relatively small surface area to volume ratio. As a result, eukaryotic cells are generally much larger than prokaryotic cells (Fig. 2.2). In the typical animal cell, the various specialized organelles occupy about half the total cell volume.

STRUCTURE AND FUNCTION OF EUKARYOTIC ORGANELLES

Eukaryotic cells have a complex ultrastructure comprising membranous and non-membranous organelles. These structures serve specific functions within the cell.

Plasma membrane

The plasma membrane is a selectively permeable barrier that surrounds the eukaryotic cell forming a dynamic interface between the cytosol and the environment. Nonpolar (lipid soluble) molecules diffuse across the lipid bilayer by passive transport, while polar molecules are transported between the cell and the extracellular fluid by proteins embedded within the bilayer (either by facilitated diffusion or active transport – see Ch. 3).

Membranous organelles

Membranous organelles are enclosed within a phospholipid bilayer. They maintain discrete biochemical environments that contain characteristic sets of enzymes.

Nucleus

The nucleus (Fig. 2.3) is bound by a double membrane, with a distinct space between which is continuous at points with the endoplasmic reticulum. The nucleus contains the genetic material of the cell (chromosomes – see Ch. 6). It may also contain one or more dense-staining areas called nucleoli, the main role of which is the biosynthesis of ribosomal RNA (rRNA) and the assembly of ribosomes.

> **HINTS AND TIPS**
>
> **Collagen – from gene to extracellular matrix**
>   Nucleus: the collagen gene is transcribed to yield mRNA, which leaves the nucleus via nuclear pores.

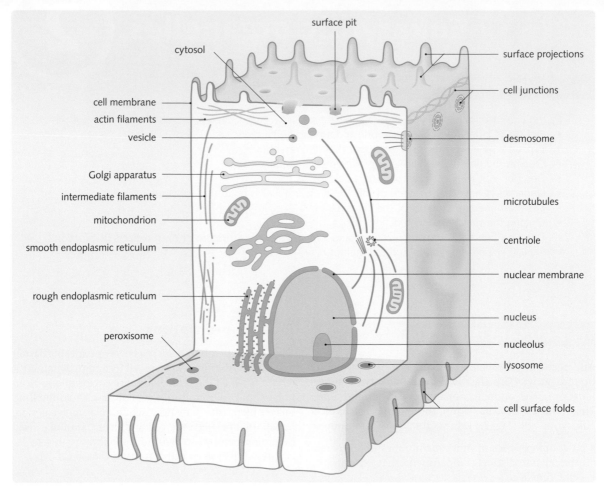

Fig. 2.1 Structure of a typical eukaryotic cell. Genetic material is contained within the nuclear space surrounded by the nuclear membrane. Membrane-bound organelles serve as compartments for specific cellular functions, permitting greater cellular specialization and diversity. Cytoskeletal components maintain cell shape and facilitate dynamic functions such as endocytosis. (Adapted from Stevens and Lowe, 1997.)

Fig. 2.2 Basic features of prokaryotic and eukaryotic cells.	
Prokaryotic cells	**Eukaryotic cells**
Includes bacteria and blue–green algae	Four major groups: Protista, fungi, plants and animals
No true nucleus	True nucleus
DNA circular and free	DNA linear and within nucleus
No membrane-bound organelles	Internal compartmentalization with organelles, hence division of labour (specialization)
Simple binary reproduction	Mitotic reproduction (and meiotic)
No development of tissues	Tissue and organ systems common
Multicellular types rare	Independent unicellular organism or part of multicellular organism
Size: 1–10 μm	Size: 10–100 μm

The nucleus is the location of a number of events, including:

- sequestration and replication of DNA
- transcription and modification of RNA
- facilitated selective exchange of molecules, such as RNA, e.g. transfer RNA (tRNA), with the cytoplasm via nuclear pores
- production of ribosomes within the nucleolus.

Mitochondria

The structure of a mitochondrion is illustrated in Figure 2.3. Mitochondria are semi-autonomous and self-replicating organelles, with their own ribosomes, RNA and several copies of a circular DNA molecule – the mitochondrial genome (see Ch. 6). Mitochondrial

DNA shows a maternal inheritance pattern, and mutations within can yield genetic disease.

The main function of mitochondria in aerobic cells is oxidative phosphorylation and the production of energy though synthesis of ATP. In addition to this crucial role, mitochondria are also important in apoptosis and cellular Ca^{2+} handling. They are also potent producers of free-radicals, a process implicated in many pathological states, including Alzheimer disease and cardiovascular disease.

Rough (granular) endoplasmic reticulum

Rough endoplasmic reticulum (RER) is a labyrinth of membranous sacs, called cisternae, to which ribosomes are attached giving a 'rough' appearance on electron

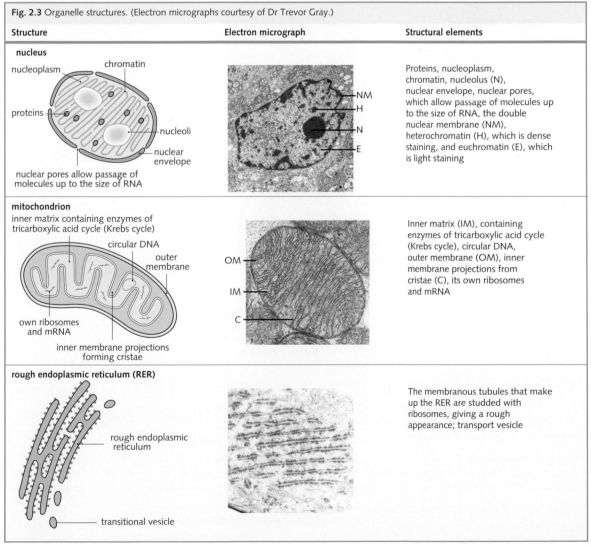

Fig. 2.3 Organelle structures. (Electron micrographs courtesy of Dr Trevor Gray.)

Structure	Electron micrograph	Structural elements
nucleus nucleoplasm, chromatin, proteins, nucleoli, nuclear envelope, nuclear pores allow passage of molecules up to the size of RNA	NM, H, N, E	Proteins, nucleoplasm, chromatin, nucleolus (N), nuclear envelope, nuclear pores, which allow passage of molecules up to the size of RNA, the double nuclear membrane (NM), heterochromatin (H), which is dense staining, and euchromatin (E), which is light staining
mitochondrion inner matrix containing enzymes of tricarboxylic acid cycle (Krebs cycle), circular DNA, outer membrane, own ribosomes and mRNA, inner membrane projections forming cristae	OM, IM, C	Inner matrix (IM), containing enzymes of tricarboxylic acid cycle (Krebs cycle), circular DNA, outer membrane (OM), inner membrane projections from cristae (C), its own ribosomes and mRNA
rough endoplasmic reticulum (RER) rough endoplasmic reticulum, transitional vesicle		The membranous tubules that make up the RER are studded with ribosomes, giving a rough appearance; transport vesicle

Continued

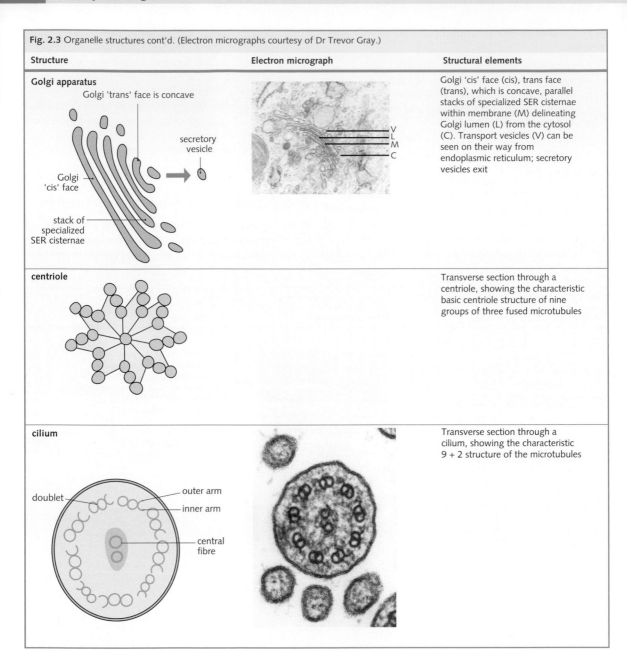

Fig. 2.3 Organelle structures cont'd. (Electron micrographs courtesy of Dr Trevor Gray.)

Structure	Electron micrograph	Structural elements
Golgi apparatus		Golgi 'cis' face (cis), trans face (trans), which is concave, parallel stacks of specialized SER cisternae within membrane (M) delineating Golgi lumen (L) from the cytosol (C). Transport vesicles (V) can be seen on their way from endoplasmic reticulum; secretory vesicles exit
centriole		Transverse section through a centriole, showing the characteristic basic centriole structure of nine groups of three fused microtubules
cilium		Transverse section through a cilium, showing the characteristic 9 + 2 structure of the microtubules

microscopy (Fig. 2.3). The cisternal ribosomes make polypeptides, which are then in turn:

- inserted into the membrane
- released into the lumen of the cisternae
- transported to the Golgi complex or elsewhere.

Proteins made within RER are kept within vesicles or secreted, and cells that make large quantities of secretory protein have large amounts of RER (e.g. pancreatic acinar cells, plasma cells).

Collagen – from gene to extracellular matrix

Rough endoplasmic reticulum (RER): the mRNA is translated to produce the immature 'preprocollagen' polypeptide on the surface of the RER. Preprocollagen is a precursor molecule, consisting of procollagen and an N-terminal 'signal peptide', required to carry the polypeptide across the membrane of the RER. Within the lumen of the RER, the signal peptide is cleaved, yielding procollagen. From here, procollagen undergoes a series of covalent modifications, including hydroxylation, to yield hydroxyproline and hydroxylysine, and glycosylation of hydroxylysine. Groups of three procollagen α-chains are arranged into a trimeric molecule and subsequently into a triple helical conformation. Molecules are then packaged in transport vesicles for their onward journey to the cis face of the Golgi.

Smooth (agranular) endoplasmic reticulum

Smooth endoplasmic reticulum (SER) is a labyrinth of cisternae with many enzymes attached to its surface or found within its cisternae. SER:

- makes steroid hormones (e.g. in the ovary)
- detoxifies body fluids (e.g. in the liver).

Clinical Note

Liver cells have very highly developed smooth endoplasmic reticulum (SER), which contains a large proportion of the body's detoxifying enzymes, including the P450 cytochromes. Enzyme induction by certain drugs may be followed by proliferation of the SER, accompanied by the production of more detoxifying enzymes, which can result in enhanced breakdown of a drug. This has an impact upon the dose required to reach therapeutic threshold and leads to a loss of therapeutic effect.

Golgi apparatus

The Golgi apparatus (Fig. 2.3) is a polarized system of membranous flattened sacs, each with a cis and a trans face. They are involved in modifying (e.g. by glycosylation), sorting and packaging macromolecules for secretion or delivery to other organelles. Protein sorting and packaging occurs at the trans face. Cells that produce many secretory products have well-developed Golgi apparatus (e.g. hepatocytes).

HINTS AND TIPS

Collagen – from gene to extracellular matrix
  Golgi apparatus: transport through the Golgi apparatus is accompanied by modification of oligosaccharide groups.

Together with the endoplasmic reticulum, the Golgi makes up the endomembrane system.

Lysosomes

Lysosomes are the primary components of intracellular digestion (see Ch. 4), and are derived from the trans face of the Golgi apparatus. Cells specializing in phagocytosis (e.g. macrophages) have many lysosomes which:

- contain granular amorphous material and about 60 types of hydrolytic enzymes
- vary in size from 50 nm to over 1 mm
- digest material with hydrolases that are active at acid pH.

Peroxisomes

Peroxisomes are vesicular bodies that are smaller than lysosomes, and contain specific enzymes. They are derived from the endoplasmic reticulum. Peroxisomal functions include biogenesis reactions (i.e. cholesterol, bile and plasmalogens), the degradation of fatty acids (by way of β-oxidation) and the breakdown of excess purines to urea. They are also required for the breakdown of toxic compounds, and so are found in high abundance in the liver and kidney. Diseases where peroxisomes have been implicated include X-linked adrenoleukodystrophy and Zellweger syndrome.

Clinical Note

Zellweger syndrome is a rare, autosomal recessive disorder characterized by the reduction or absence of peroxisomes, which leads to multiple disturbances of lipid metabolism due to a build up of very long chain fatty acids. It also leads to severe neurological dysfunction, with indicates that perioxisomes are required for normal central nervous system neuronal migration. Death usually occurs within 6 months of onset, usually as a result of respiratory distress, gastrointestinal bleeding or liver failure, highlighting the physiological importance of these organelles.

Secretory vesicles

Secretory vesicles are organelles that deliver their contents, such as hormones and neurotransmitters, to the outside of the cell by fusing with the plasma membrane. They are derived from the trans face of the Golgi apparatus and their production may be:

- constitutive secretion (e.g. collagen from fibroblasts and albumin from hepatocytes)
- regulated release (e.g. insulin from b-cells of pancreatic islets).

Collagen – from gene to extracellular matrix

Secretory vesicle: modified procollagen is secreted constitutively by fibroblasts. Upon secretion, uncoiled terminal ends of procollagen are cleaved to form tropocollagen, which themselves aggregate with the crosslinking of lysine and hydroxylysine residues are crosslinked to form a collagen fibril.

Non-membranous organelles

Ribosomes

Eukaryotic cells possess '80 S' ribosomes, consisting of a small 40 S subunit and a large 60 S subunit. Ribosomes that synthesize proteins for use within the cytosol of the cell are found suspended within in the cytosol, whereas proteins destined for the plasma membrane or cell vesicles are attached to the cytosolic face of the membranes of the endoplasmic reticulum (see pp. 15–17).

The cytoskeleton

The cytoskeleton is the internal framework of the cell, consisting of filaments and tubules. Cytoskeletal structures maintain and change cell shape by rearrangement of the cytoskeletal elements. They are essential for endocytosis, cell division, amoeboid movements, and contraction of muscle cells. There are several classes of cytoskeletal structural components:

- microfilaments formed from actin
- microtubules formed from tubulin
- intermediate filaments formed from intermediate filament proteins, such as cytokeratin.

These structures may be cross-linked by other proteins into networks or specialized organelles, the most common of which are as follows:

- Centrioles – these usually occur in pairs, which in non-dividing cells are aligned at right angles to each other. The basic centriole structure is one of nine groups of three fused microtubules arranged as a cylinder around a central cavity (Fig. 2.3). As cells prepare to divide, the centriole pairs separate and go to opposite ends of the cell, where they act as the site of spindle assembly in cell division.
- Cilia – used by some cells to aid the movement of a cell or substance over the surface of cells (e.g. fallopian cells move ova towards the uterus). They are attached to structures known as basal bodies, identical in structure to centrioles, which are anchored to the cytoplasmic side of the plasma membrane. Microtubules, arising from the basal bodies, are arranged in a '9 + 2' arrangement consisting of nine microtubule doublets

surrounding two single microtubules (Fig. 2.3). Dynein side arms extend between adjacent doublets and hydrolyse ATP to generate a sliding force between them. This action underlies ciliary beating by bending of these arms. Cilia can also be non-motile (sensory), as seen in the rod cells of the eye and the terminal fibres of olfactory neurons, where they have a '9 + 0' arrangement, with the central pair of microtubules being absent.

Clinical Note

Kartagener syndrome is an autosomal recessive syndrome typified by situs inversus, chronic sinusitis and bronchiectasis. Cilia motility is produced by dyneins and other related microtubule proteins, ultrastructural defects leading to ciliary diskinesia which have been hypothesized as being at the root of Kartagener syndrome. Impaired ciliary function leads to reduced or absent mucus clearance in the lungs, and susceptibility to chronic, recurrent respiratory infections. Situs inversus (a mirror image arrangement of the organs) is thought to arise as interplay between motile and sensory cilia is required for determination of left–right axis in early vertebrate development.

- Flagella – very long cilia used for propulsion by spermatozoa. Eukaryotic flagella have a '9 + 2' structure and, therefore, have both a different structure and origin to prokaryotic flagella.
- Microvilli – non-motile extensions of plasma membrane supported by actin, which increase the surface area of the cell (e.g. the small intestine brush-border).
- Pseudopodia – although not true specializations, pseudopodia are extensions of the plasma membrane, formed by actin polarization. They are commonly seen in phagocytes, such as macrophages.
- Junctions – these are points of adhesion between cells and other cells, and between cells and their basement membrane. They are discussed in more detail in Chapter 4.

CELL DIVERSITY IN MULTICELLULAR ORGANISMS

Cell specialization

It is thought that multicellular organisms evolved as a result of specialized cells acting together and combining to form a single organism, able to exploit ecological niches not available to any of its component cells acting alone.

Similar types of specialized cells combine together to form tissues, of which there are four main types each adapted to a specific function (Fig. 2.4). By

Fig. 2.4 Cell specialization in epithelial, connective, nervous and muscle tissue.

Tissue structure	Function	Specialized cell types
Epithelial tissue Consists of continuous sheets of cells that are bound together by tight junctions	Epithelial tissue lines inner and outer surfaces of the body to form a selectively permeable barrier Epithelial surfaces may be specialized for: • absorption • substance movement • secretion	Absorptive cells—the luminal plasma membrane is folded into microvilli to increase the surface area (e.g. intestinal villi) Ciliated cells—the luminal plasma membrane is coated with cilia that beat in synchrony (e.g. tracheal mucociliary escalator) Secretory cells—the RER and Golgi apparatus are highly developed (e.g. chief cells in the stomach)
Muscle tissue Consists of groups of cells containing fibrillar proteins arranged in an organized manner in the cytoplasm and linked by intermolecular bonds Skeletal and cardiac muscle appear striated	Muscle functions to produce movement. Contraction results from the rearrangement of internal bonds between fibrillar proteins Muscle tissue is specialized to allow: • voluntary movement of the skeleton • involuntary movement of substances through the viscera • continuous synchronous contraction of the heart	Skeletal muscle cells—each muscle fibre is an enormous multinucleated cell that extends the full length of the muscle (nuclei are located at the cell periphery). Thus, excitation results in simultaneous contraction of the full length of muscle in a longitudinal direction Visceral muscle cells—cells are relatively small with tapered ends and only a single nucleus. The cells are arranged in layers at right angles to one another to facilitate peristalsis Cardiac muscle cells—cells are Y shaped with one nucleus that is centrally located. The longitudinal branches of adjacent cells join at intercalated discs. This structure allows for the rapid spread of contractile stimuli from one cell to another
Connective tissue Consists of cells and extracellular material. The extracellular material is secreted by the cells and determines the physical properties of the tissue. It is composed of ground substance, fibres (collagen and elastin), and structural glycoproteins	Connective tissue provides structural and metabolic support for other tissues and organs Loose connective tissue acts as biological packing material Dense connective tissue provides tough physical support. For example, it forms the skeleton, the dermis and organ capsules	Fibroblasts—synthesize and maintain extracellular material. They are active in wound healing where specialized contractile fibroblasts (myofibroblasts) bring about shrinkage of scar tissue Adipocytes—store and maintain fat. They are found in clumps in loose connective tissue and form the main cell type in adipose tissue. Fat stored in adipocytes forms a large droplet that occupies most of the cytoplasm Chondroblasts and chodrocytes—produce and maintain cartilage Osteoblasts, osteocytes and osteoclasts—specialized cells that produce, maintain and break down bone, respectively Blood cells—leukocytes (white blood cells), platelets (thrombocytes), and erythrocytes (red blood cells). Functions include immune defence, blood clotting and oxygen carriage, respectively
Nervous tissue Peripheral nervous tissue is composed of neurons and Schwann cells Central nervous tissue consists of neurons and neuroglial cells (oligodendrocytes, astrocytes, microglia, and ependymal cells). It is divided macroscopically into grey and white matter. White matter consists of tracts of myelinated nerves. Grey matter contains neuronal cell bodies	Nervous tissue detects changes in the internal and external environments. By transmitting and processing this information it coordinates the activities of the multicellular organism to produce an appropriate response Nervous tissue is specialized for: • sensing environmental change • conducting information • integrating and analysing information	Sensory receptors—there are numerous cell types specialized for detecting environmental change, for example, Pacinian corpuscles (mechanoreceptor that detects skin pressure) Neurons—these are specialized for receiving and transmitting information. Therefore, they synthesize neurotransmitters and neurotransmitter receptors. Multiple dendrites allow communication with many neighbouring cells and function as sites of information input. The axon may be extremely long, and it facilitates transmission of information to distant sites. Terminal boutons arise at the end of the axon and communicate with other nerve cells or the effector organ Schwann cells/oligodendrocytes—specialist cells that wrap around the neuronal axon to form the myelin sheath and provide structural and metabolic support

definition, specialized cells show structural features that enable them to perform their designated function. In order to cooperate and coordinate their activities in the multicellular animal:

- cells are bound together by adhesions between their plasma membranes and the extracellular matrix (see Ch. 4)
- cells interact and communicate with one another (see Ch. 3).

Differentiation

Over 200 types of cell are identifiable in human tissue, differing in terms of their structure, function and chemical metabolism. All cell types are derived from a single cell (the zygote) following conception. The zygote is described as being totipotent since it is ultimately able to differentiate into all the cell types that make up the adult organism. As there is no loss of genetic material from somatic cells during human development (erythrocytes are an exception), different cell types arise as a result of mechanisms such as imprinting and differential gene expression.

During development, cell differentiation is driven by successive cascades of proteins that regulate the DNA in each cell, restricting transcription from specific sections of the genome, and promoting it in others. The developmental signals that initiate differentiation in a human cell come from the cells that surround it. However, even when removed from its normal environment the differentiated cell and its progeny will retain many of its functional characteristics. This concept is termed cell memory due to the transmission of epigenetic changes to daughter cells.

Stem cells play a key role in generating and maintaining this diversity. Embryonic stem cells have the ability to develop into any type of fully developed cell in the body; they are pluripotent. Adult stem cells replenish the stock of some cell types to maintain adequate tissue function over a lifetime. These particular stem cells are multipotent, as in their specific tissue types they can give rise to many types of cells. Induced pluripotent stem cells are adult stem cells which have been artificially reprogrammed to behave like embryonic stem cells.

The cell membrane ③

● Objectives

By the end of this chapter you should be able to:

- Draw a labelled diagram of the fluid mosaic model of the plasma membrane.
- Discuss the chemical properties of phospholipids and their significance in the plasma membrane.
- Describe the modes of movement available to phospholipids and proteins in the plasma membrane.
- Define diffusion, osmosis, osmotic pressure, isotonic, hypotonic and hypertonic.
- Describe the relative concentrations of K^+, Na^+, Cl^- and Ca^{2+} across the resting cell membrane.
- Understand the differences between facilitated, primary active and secondary active transport.
- Outline the structure of the Na^+/K^+ ATPase, define what type of transport it mediates and describe its function with reference to ionic gradients.
- Define the terms endocrine, paracrine and autocrine in relation to cell signalling.
- Understand the process of signal transduction and the classes of molecules involved.
- Understand how steroid hormone receptors differ from cell surface receptors.

STRUCTURE OF THE CELL MEMBRANE

Fluid mosaic model

The fluid mosaic model, first proposed in 1972, is one of a biological membrane consisting of a phospholipid bilayer with proteins embedded in it (Fig. 3.1). The model has been verified by both freeze fracture and freeze etching electron microscopy. The phospholipid molecules, the major component, are amphipathic, i.e. have hydrophobic and hydrophilic regions (Fig. 3.2). They form a stable bilayer in aqueous solutions due to:

- hydrophilic interactions of polar head groups with the extracellular and intracellular aqueous environments
- hydrophobic interactions of the fatty acid molecules in the bilayer interior.

The membrane is a dynamic structure and many proteins are able to move freely through it like 'icebergs floating in a sea of phospholipids'.

Under normal cellular conditions, the turnover of the plasma membrane is dependent upon the delivery of new membrane components to the membrane in late endosomes and lysosomes.

HINTS AND TIPS

The fluid mosaic model is a common topic in exams; remember to draw a diagram including peripheral and integral proteins, even for short-answer questions. Mention that the membrane is dynamic, and in essay questions discuss the movement of phospholipids and proteins within it.

Components of the biological membrane

Lipids

A lipid is a molecule that is soluble in an organic solvent (e.g. chloroform), but only sparingly soluble in water. Triglycerides, which are hydrophobic, consist of three fatty acids attached to a glycerol backbone by ester linkages.

Phospholipids

The hydrophilic moiety in a phospholipid is due to the substitution of one of the fatty acid chains for an amine-containing polar group linked to glycerol (at C3) by a phosphodiester bond (see Fig. 3.2). There are four plasma membrane phospholipids (Fig. 3.3):

1. Phosphatidylethanolamine
2. Phosphatidylserine
3. Phosphatidylcholine
4. Sphingomyelin (not a true phospholipid as it possesses an acylated sphingosine [ceramide] not a glycerol back bone).

Phospholipids are arranged asymmetrically, with almost all of the phosphatidylcholine and sphingomyelin

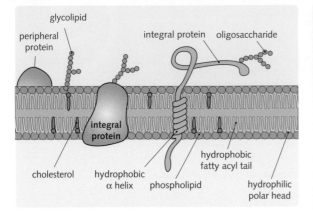

Fig. 3.1 Cell membrane fluid mosaic model.

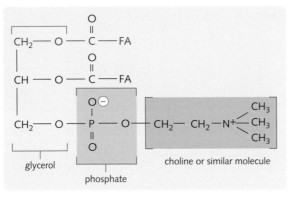

Fig. 3.2 Structure of a phospholipid. The shaded areas correspond to the hydrophilic parts of the molecule. FA, fatty acid.

occuring in the outer monolayer, while phosphatidylethanolamine and phosphatidylserine occur predominantly in the inner monolayer. Since they aggregate into a continuous sheet impervious to ions, the molecules must pack very closely. Most lipids pack as cylinders or slightly truncated cones, although lysophospholipids, intermediates formed during digestion of dietary and biliary phospholipids and which lack one fatty-acid chain, are shaped like cones (Fig. 3.4). The presence of a double bond in a fatty-acid side chain introduces a kink, which disrupts van der Waals forces and reduces the ability of the molecule to fit tightly with its neighbours, increasing membrane fluidity.

> **Clinical Note**
>
> The plasma membrane is a dynamic structure, with a constant turnover of constituents. Niemann–Pick (NP) disease is a collection of lysosomal storage diseases

displaying an autosomal recessive inheritance pattern. Types A and B result from a deficiency in lysosomal acid sphingomyelinase required to hydrolyse spent membrane sphingomyelin to yield ceramide and phosphocholine, leading to its accumulation in reticuloendothelial foam cells within the spleen, liver, lungs, bone marrow and brain.

Cholesterol

Cholesterol is a lipid molecule consisting of four hydrophobic rings and a hydrophilic hydroxyl group (Fig. 3.5). It orientates in the membrane such that the rings lie parallel to the hydrophobic fatty-acid groups, with the hydroxyl group forming a hydrogen bond with the carboxyl group on an adjacent phospholipid. The net result is that:

- at physiological temperature, cholesterol restricts the movement of the fatty-acid chains and stabilizes the membrane by reducing fluidity
- at low temperatures it inhibits phospholipid packing, which increases membrane fluidity.

Membrane proteins

Integral proteins

Integral proteins span the membrane and they have intracellular and extracellular domains.

- The membrane-spanning domains are rich in hydrophobic amino acid residues and traverse the membrane as α-helical loops.
- The cytosolic and extracellular domains are rich in polar amino acid residues.
- Extracellular domains may be glycosylated.

Monotopic integral proteins traverse the membrane once, while bitopic and polytopic proteins pass through the membrane twice and many times, respectively.

> **Clinical Note**
>
> Polycystin 2, an integral membrane protein encoded by the PKD2 gene, is required for normal tubulogenesis in the kidney. It is found to be mutated in approximately 15% of all cases of autosomal dominant polycystic kidney disease (ADPKD). Normally, polycystin 2 interacts with polycystin 1 to act as a calcium permeable cation channel, and mutations in either of these proteins cause virtually indistinguishable clinical presentations. Defective polycystins appear to affect epithelial cell maturation, resulting in the development of cysts of varying sizes throughout the cortex and medulla.

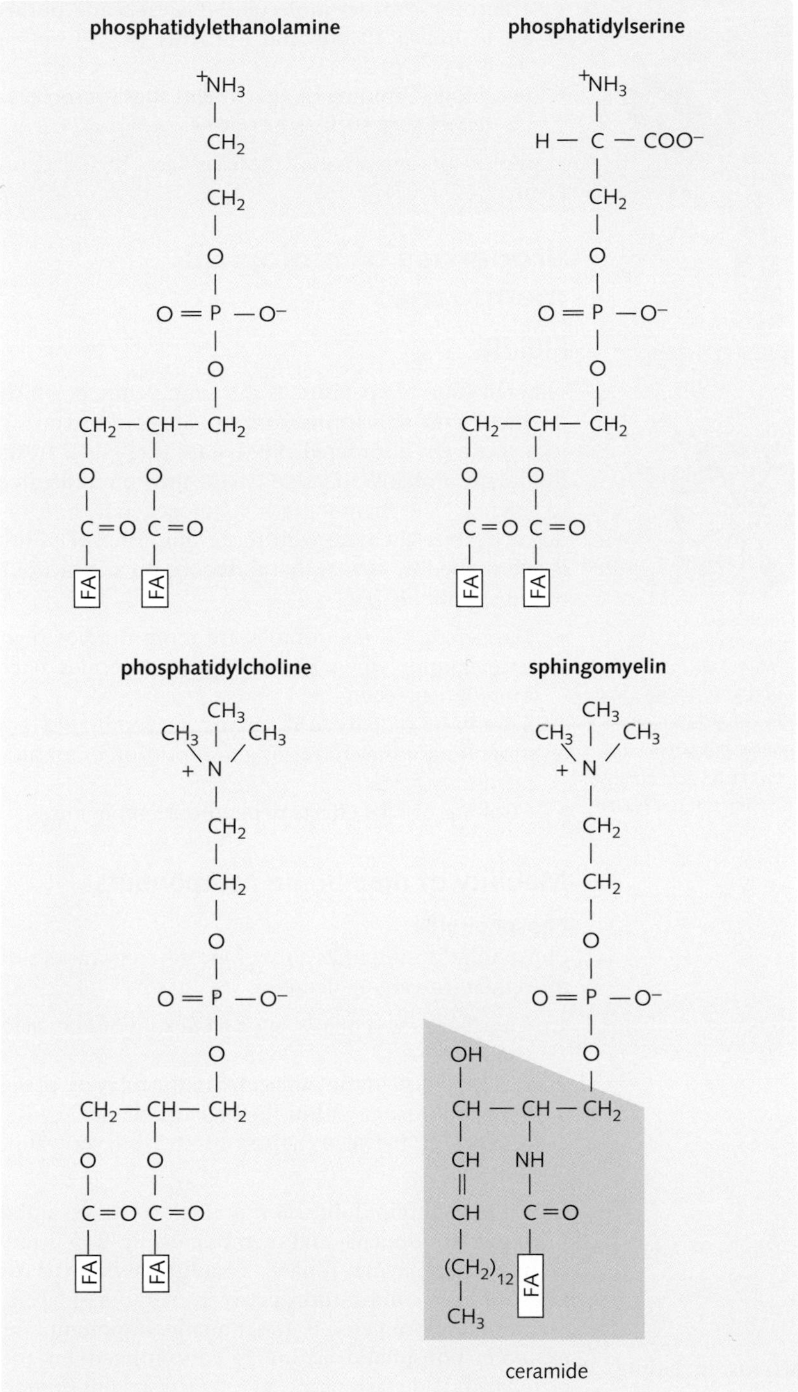

phosphatidylethanolamine

$$^+NH_3$$
$$|$$
$$CH_2$$
$$|$$
$$CH_2$$
$$|$$
$$O$$
$$|$$
$$O = P - O^-$$
$$|$$
$$O$$
$$|$$
$$CH_2 - CH - CH_2$$
$$|\quad\quad\;|$$
$$O\quad\quad O$$
$$|\quad\quad\;|$$
$$C = O\;\; C = O$$
$$|\quad\quad\;|$$
[FA] [FA]

phosphatidylserine

$$^+NH_3$$
$$|$$
$$H - C - COO^-$$
$$|$$
$$CH_2$$
$$|$$
$$O$$
$$|$$
$$O = P - O^-$$
$$|$$
$$O$$
$$|$$
$$CH_2 - CH - CH_2$$
$$|\quad\quad\;|$$
$$O\quad\quad O$$
$$|\quad\quad\;|$$
$$C = O\;\; C = O$$
$$|\quad\quad\;|$$
[FA] [FA]

phosphatidylcholine

$$CH_3$$
$$CH_3\;|\;CH_3$$
$$_+\;N$$
$$|$$
$$CH_2$$
$$|$$
$$CH_2$$
$$|$$
$$O$$
$$|$$
$$O = P - O^-$$
$$|$$
$$O$$
$$|$$
$$CH_2 - CH - CH_2$$
$$|\quad\quad\;|$$
$$O\quad\quad O$$
$$|\quad\quad\;|$$
$$C = O\;\; C = O$$
$$|\quad\quad\;|$$
[FA] [FA]

sphingomyelin

$$CH_3$$
$$CH_3\;|\;CH_3$$
$$_+\;N$$
$$|$$
$$CH_2$$
$$|$$
$$CH_2$$
$$|$$
$$O$$
$$|$$
$$O = P - O^-$$
$$|$$

$$OH\quad\quad O$$
$$|\quad\quad\;|$$
$$CH - CH - CH_2$$
$$|\quad\quad\;|$$
$$CH\quad NH$$
$$\|\quad\quad\;|$$
$$CH\quad C = O$$
$$|\quad\quad\;|$$
$$(CH_2)_{12}\;[FA]$$
$$|$$
$$CH_3$$

ceramide

Fig. 3.3 The three major phospholipids, phosphatidylethanolamine, phosphatidylserine and phosphatidylcholine, have different polar head groups. FA, fatty acid.

Peripheral proteins

Peripheral proteins are associated with either the cytoplasmic or extracellular leaf of the lipid bilayer. They may be attached to the membrane:

- electrostatically – integral proteins
- covalently – non-protein cytoplasmic layer components
- covalently – phospholipids via an extracellular oligosaccharide linker.

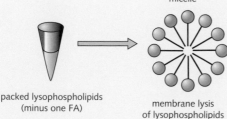

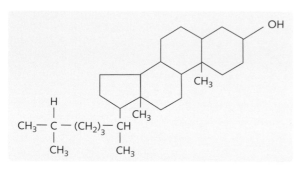

Fig. 3.4 Packing of phospholipids. The nature of the fatty-acid side chains influences their packing. Lysophospholipids, which lack one fatty acid, form micelles preferentially. However, the favoured structure for phospholipids with two fatty-acid chains in aqueous solution is a lipid bilayer because they are too bulky to form micelles. FA, fatty acid.

Fig. 3.5 Structure of cholesterol.

Functions of membrane proteins

Membrane proteins have several functions, including:

- Markers – the carbohydrate chains of glycoproteins aid in self-recognition.
- Enzymes – water-soluble enzymes associate with the polar heads of membrane phospholipids (e.g. phospholipases).
- Anchors – membrane proteins act as attachment points to the cytoskeleton and extracellular matrix, aiding strength and adhesion.

- Transport – carrier molecules, channels and porins act to transport ions and nutrients in and out of the cell (see p. 27).
- Receptors – binding or attachment sites for molecular messengers, such as hormones (see p. 32).

Examples of membrane proteins are included in Figure 3.6.

Properties of biological membranes

Fluidity

The transition temperature is the temperature at which the membrane transforms from a rigid gel-like structure to a relatively disordered, fluid state (Fig. 3.7). In its fluid state, proteins embedded in the membrane are free to interact. The membrane is heterogeneous, and ordered regions alternate with more fluid ones. Fluidity is determined by factors that influence the interactions of phospholipids (Fig. 3.8).

- Temperature – membranes are more fluid at high temperatures when phospholipid molecules have more kinetic energy.
- Saturation of fatty-acid chains – membranes are more fluid if they have a high proportion of unsaturated fatty acids.
- Cholesterol – its effects depend on temperature.

Mobility of membrane components

Phospholipids

Phospholipid molecules show four different modes of movement to varying degrees:

1. Intrachain movements, such as flexing of fatty-acid chains
2. Axial rotation of the molecules in the bilayers' plane
3. Lateral diffusion within the bilayers' plane
4. Movement between halves of the bilayer ('flip-flop').

Lateral phospholipid diffusion occurs readily, resulting in a fluid two-dimensional membrane. Flip-flop is rare without the enzyme 'flipase', enabling asymmetry of phospholipid composition between membrane layers. Membrane asymmetry is functionally important, for example, phosphatidylserine is concentrated on the cytoplasmic side facilitating its interaction with protein kinase C.

Proteins

Proteins show three types of movement:

1. Lateral – this may be restricted by cytoskeletal interactions or covalent attachment to lipid molecules
2. Axial rotational – in the plane of the membrane
3. Conformational changes.

Fig. 3.6 Examples of membrane proteins and their functions. CFTR, cystic fibrosis transmembrane regulator.

Protein	Type	Bonding with membrane	Function
Cadherin	Monotopic integral	Hydrophobic with phospholipids	Mediates cell–cell adhesion
CFTR	Polytopic integral	Hydrophobic with phospholipids	Gated chloride channel in epithelial tissue
Ankyrin	Peripheral	Electrostatic with the anion exchange protein on the cytoplasmic surface of the lipid bilayer	Maintains erythrocyte structure by forming a link between spectrin and the anion exchange protein (band 3)
Ras	Peripheral	Covalent attachment to the cytoplasmic layer of the lipid bilayer	GTP-binding protein that relays signals from the cell surface to the nucleus

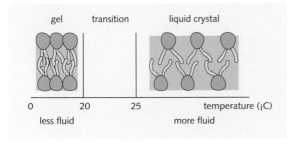

Fig. 3.7 Membrane transition from gel to liquid crystal.

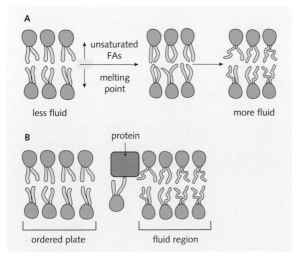

Fig. 3.8 Factors affecting membrane fluidity. (A) Increasing unsaturated fatty acid (FA) concentration decreases the melting point and increases fluidity. (B) Uneven distribution of membrane lipids.

Flip-flop is thermodynamically unfavourable, as it requires hydrophilic protein moieties to rotate through the bilayer and is, therefore, rare. Protein directionality is important for function, and ensures, for example, correct orientation of cell surface receptors.

Permeability

There are three main forms of transport across the membrane:

1. Passive diffusion
2. Facilitated diffusion
3. Active transport.

Though permeable to lipid-soluble compounds, lipid bilayers are impermeable to ionic and polar substances, which require dedicated channels to cross the membrane. The exception is water, which, although polar, is membrane permeable due to its small size.

TRANSPORT ACROSS THE CELL MEMBRANE

Concepts

Concentrations are measured in moles per litre and the dissociation of ions is not taken into account.

$$1 \text{ mol} = 6.02 \times 10^{23} \text{ molecules (of any kind)}$$
$$= \text{Avogadro constant (the number of atoms in exactly 12 g of carbon 12)}$$

For example molar solutions of:

- NaCl contains 1 mol of Na^+ atoms and 1 mol of Cl^- atoms in 1 L.

- Sucrose contains 1 mol of sucrose in 1 L.
- $CaCl_2$ contains 1 mol of Ca^{2+} and 2 mol of Cl^- in 1 L.

Diffusion is the movement of particles from a region of high concentration to a region of low concentration until evenly distributed. Osmosis is the movement of solvent molecules (usually water) across a semi-permeable membrane from a region of high solvent concentration to a region of low solvent concentration. The osmotic pressure is the pressure required to prevent the net movement of pure water into an aqueous solution across a semi-permeable membrane. In osmotic pressure the dissociation of ions is important.

> 1 osmol = the amount of substance that dissociates in solution to form 1 mole of osmotically active particles

For example (and assuming full dissociation):

- 1 Osmol/L of NaCl contains 0.5 mol of the ion Na^+ and 0.5 mol of the ion of Cl^- per litre.
- 1 Osmol/L of sucrose contains 1 mol of sucrose per litre.
- 3 Osmol/L of $CaCl_2$ contains 1 mol of the ion Ca^{2+} and 2 mol of the ion Cl^- per litre.

The osmolarity of plasma is critical, as changes affect plasma volume, cell volume, and water and ion homeostasis. The range for normal plasma osmolarity is 280–295 mosmol/L. Serum plasma values are often given as osmolality: the concentration of osmotically active particles in solution per kilogram of solvent (osmol/kg). Dissociation is affected by pH, temperature and binding of ions to compounds (e.g. Ca^{2+} binding to myosin during muscle contraction). In living systems, K^+, Na^+, and Cl^- are fully dissociated, whereas Ca^{2+}, Mg^{2+} and H^+ are only partially dissociated.

> **Clinical Note**
>
> Dehydration leads to a relative increase in serum solute load and, therefore, an increase in osmolality. Detected in the hypothalamus, anti-diuretic hormone (ADH) release from the anterior pituitary is triggered. ADH increases the permeability of the distal tubules and collecting ducts of the kidney, reducing the amount of water excreted and leading to a relative decrease in serum solute load and a lowering of serum osmolality. If the osmolality of the blood plasma becomes too low, then the output of the kidney is enhanced by the release of atrial natriuretic peptide (ANP), returning the solute load to the normal range.

The term 'tonicity' relates to the behaviour of cells immersed in a solution. It is the effective osmolality and is equal to the sum of the concentrations of the solutes that have the capacity to exert an osmotic force across the membrane.

- Isotonic extracellular solutions have the same osmotic pressure as the inside of the cell, so osmosis does not occur and the cell remains the same size.
- Hypotonic solutions are less concentrated, so water will pass into the cell and it will swell.
- Hypertonic solutions are more concentrated, so water will pass out of the cell and it will shrink.

Distribution of ions across the cell membrane

Life's essential chemical reactions occur within narrow physiological parameters, so the cell must regulate the entry and exit of intracellular molecules. Some biological processes, such as muscle contraction, depend upon an electrochemical gradient across the cell membrane. The distribution of ions across the cell membrane is shown in Figure 3.9. Distribution is influenced by:

- the semi-permeable membrane concept
- electrochemical gradient
- pumps.

Semi-permeable membrane concept

Polar molecules cannot diffuse through the lipid bilayer, and they rely on the proteins embedded in the membrane for transport. These proteins are generally specific for particular molecules. The cell regulates the activity of membrane transport proteins such that only certain molecules can get through at any one time.

Electrochemical gradient

It is thermodynamically favourable for ions to move from areas of high concentration to low concentration and for positively charged ions to move to negatively charged environments.

Fig. 3.9 Distribution of ions across the cell membrane.		
Component	**Outside**	**Inside**
K^+ (mmol/L)	4.5	140 (varies with cell type)
Na^+ (mmol/L)	140	10
Ca^{2+} total (mmol/L)	3	1
Ca^{2+} free (µmol/L)	1	0.1
Cl^- (mmol/L)	110	3
HCO_3^- (mmol/L)	24	10
pH	7.35	7
Amino acids, proteins	10	120

Pumps

Pumps are used to maintain energetically unfavourable concentration gradients. For example, the cell is able to maintain an energetically unfavourable sodium gradient because it expends energy in the form of ATP to drive sodium out of the cell.

Transport across the membrane

Transport across a biological membrane is summarized in Figure 3.10.

Passive (simple) diffusion

Passive diffusion is the free movement of molecules across a membrane down a concentration gradient. Small non-polar molecules (e.g. O_2 and CO_2) and uncharged polar molecules (e.g. urea) may diffuse directly through the lipid bilayer by this means. No energy is required and diffusion continues until equilibrium is reached. Saturation does not occur because no binding

sites are involved. The diffusion rate is directly proportional to the ion gradient, hydrostatic pressure and electrical potential, and is summarized by Fick law of diffusion:

$$\text{Rate of diffusion} = D \times \text{area} \times \Delta \text{ conc}$$

where D = diffusion constant, A = membrane area and Δ conc = concentration gradient.

Facilitated diffusion

As charged molecules cannot diffuse directly through the lipid bilayer, they depend on specific proteins. The transport of molecules by a protein receptor down a concentration gradient is called facilitated diffusion, which is a form of passive transport that continues until equilibrium is reached. Proteins mediating facilitated diffusion may be channels or carrier proteins. Transport with carrier proteins shows Michaelis–Menten kinetics (Fig. 3.11):

- substrate specificity or selectivity, affinity for a particular ligand (measured as K_L)
- saturability of ligand binding (B_{max})
- transferability (T_{max}), the maximum rate of molecule transfer across the membrane
- inhibition (e.g. transport of glucose into erythrocytes).

The rate of movement of ions through a membrane via channels depends on the concentration gradient, the speed with which the ion moves through the channel (a constant) and the number of open channels. It is, therefore, analogous to passive diffusion, with the number of open channels being equivalent to the surface area. Channels may be gated such that the cell controls when they are open. Irrespective of the electrochemical

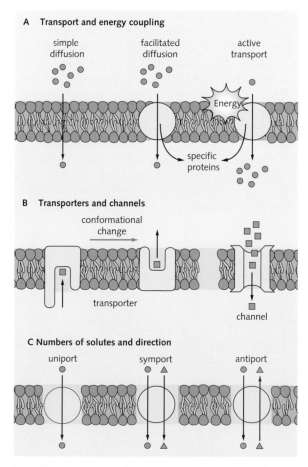

A Transport and energy coupling

simple diffusion   facilitated diffusion   active transport

Energy

specific proteins

B Transporters and channels

conformational change

transporter

channel

C Numbers of solutes and direction

uniport   symport   antiport

Fig. 3.10 A summary of solute movement across membranes. (Adapted from Baynes and Dominiczak, 1999.)

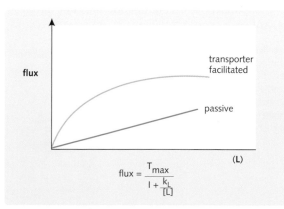

flux

transporter facilitated

passive

(L)

$$\text{flux} = \frac{T_{max}}{1 + \dfrac{k_L}{[L]}}$$

Fig. 3.11 Kinetics of facilitated diffusion. Passive diffusion is slower than transporter facilitated diffusion, its rate is directly proportional to substrate concentration. Transporter facilitated diffusion behaves like an enzyme and becomes saturated. T_{max}, maximum transport rate; k_L, affinity for ligand; $[L]$, concentration of ligand.

gradient, an ion cannot cross the membrane if there are no open channels.

Active transport

Active transport couples the movement of molecules against an unfavourable electrochemical gradient to a thermodynamically favourable reaction.

- Primary active transport is coupled directly to the hydrolysis of ATP.
- Secondary active transport is coupled indirectly to the hydrolysis of ATP.

Primary active transport

Primary active transport directly uses energy to transport molecules across a membrane. Sodium and potassium are examples of ions that are transported across the cell by primary active transport via the Na^+/K^+ dependent ATPase. For every ATP hydrolysed, this transporter pumps three Na^+ ions outward and two K^+ ions inward, against their respective concentration gradients (see p. 31).

Secondary active transport

Secondary active transport does not use ATP directly, but takes advantage of a separate existing concentration gradient. The action of the Na^+/K^+ ATPase establishes K^+ and Na^+ concentration gradients across the membrane. Movement of Na^+ into the cell, down its electrochemical gradient, is thermodynamically favoured, and in secondary transport this is coupled to the movement of a second ion against its gradient. The Na^+ electrochemical gradient may drive the transport of ions in either direction across the membrane (Fig. 3.12).

- Symports transport both ions in the same direction.
- Antiports transport the ions in opposite directions.

Transport mechanisms

Membrane transport proteins may be channels or carriers (see Fig. 3.10). Most transport proteins are reversible and, depending on the prevailing conditions, may transport ions into or out of the cell.

Ion channels

Ion channels are proteins that span the membrane and have central water-filled pores. The pores are specific, allowing either cations or anions through. Transport speed is greater than 10^6 ions/s, and it is always down a concentration gradient. Potassium channels are the most common type. One type is perpetually open, with the leakage of K^+ through these channels being critical to the membrane potential. Defects or damage can cause muscular dysfunction (e.g. periodic paralysis). Many channels are 'gated', and open and close under specific conditions.

Clinical Note

Cystic fibrosis is an autosomal recessive disease characterized by chronic lung disease, exocrine pancreatic insufficiency and male infertility. It results from mutations in the cystic fibrosis transmembrane conductance regulator (CFTR) gene, the product of which encodes a cAMP regulated gated ion channel. The channel is primarily responsible for controlling the movement of chloride from outside the cell into the cell. In the absence of a functional CFTR gene product, transport of chloride ions across apical epithelial cell membranes is impaired, leading to an accumulation of chloride outside the cell and of water and sodium in the cell, causing mucus or watery secretions outside the cells to be too thick.

Fig. 3.12 Secondary active transport. The action of the Na^+/K^+ ATPase results in a sodium concentration gradient across the membrane. This potential energy may be used to drive molecules into the cell via symports or out of the cell via antiports.

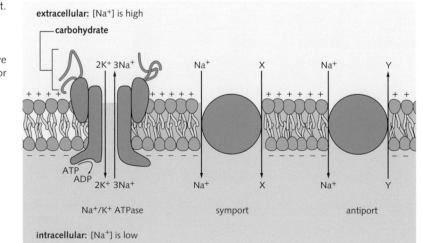

Carrier proteins

Carrier proteins bind specific ligands (the transported molecule) and undergo conformational change during transport. They transport polar and ionic molecules by active transport and facilitated diffusion. Carrier protein mediated transport is hundreds of times slower than that via ion channels, and they can become saturated, limiting the rate of transport. Uniports transport single molecules across the membrane (see Fig. 3.10). Coupled transporters transfer molecules across the membrane with simultaneous transfer of another molecule (symports and antiports). Different cells have different carrier protein populations, and so they have different permeabilities.

Glucose transporter

Most cells transport glucose by facilitated diffusion through uniports, as the concentration of glucose is greater outside the cell. However, in the intestine and kidney some cells absorb glucose from low extracellular concentrations, mediated by secondary active transport via symports co-transporting sodium.

Active transporters

Active transporters are carrier proteins that are linked to a source of energy, such as ATP or an ionic gradient.

The Na^+/K^+ ATPase pump

The sodium pump is an example of an active transporter. It is a heterodimer consisting of an α-subunit and a glycosylated β-subunit. The glycosylated subunit is important for the assembly and localization of the pump. The α-subunit is the catalytic unit, and it has binding sites for sodium and ATP on its intracellular surface, and potassium on its extracellular surface. Binding of sodium causes phosphorylation of the cytoplasmic side and a conformational change, which transfers the sodium outside the cell. Binding of potassium causes dephosphorylation, so the subunit returns to its original state, transferring the potassium inside the cell simultaneously (Fig. 3.13). The pump has several functions:

- maintaining a low intracellular sodium concentration
- maintaining a constant cell volume
- providing a sodium gradient as an energy source for co-transport, which is exploited by many processes, including the transporters that regulate intracellular pH and those that drive glucose into kidney cells
- generation of membrane potential.

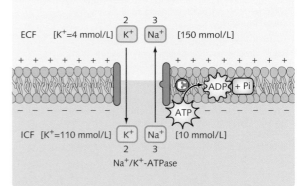

Sodium–potassium pump

Fig. 3.13 The Na^+/K^+ ATPase, associated with the hydrolysis of one molecule of ATP. ECF, extracellular fluid; ICF, intracellular fluid. (Adapted from Baynes and Dominiczak, 1999.)

MEMBRANE POTENTIAL

Definition

A membrane potential (E_m) is defined by the difference in electrical charge on each side of a membrane. It is critical in the functioning of excitable cells, especially nerve and muscle cells. These cells use the controlled opening of gated ion channels to cause a change in their membrane potential. There are three major types of gated channels in excitable cells:

1. Voltage-gated (e.g. voltage-gated sodium channels used in action potential generation)
2. Chemically-gated (e.g. acetylcholine receptor channels in neuromuscular transmission)
3. Mechanical receptors (e.g. touch receptor channels in sensory neurons).

Maintenance of membrane potential

Electrochemical potential difference of ions

Ions have an electric charge and a chemical concentration. When a solution is not at equilibrium, the movement of its ions is influenced by both these gradients. If these factors operate in different directions across a cell membrane, net ion flow will tend to be down whichever gradient is the steepest. Since the electrochemical potential difference (Δμ) of ions is defined

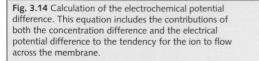

Fig. 3.14 Calculation of the electrochemical potential difference. This equation includes the contributions of both the concentration difference and the electrical potential difference to the tendency for the ion to flow across the membrane.

The electrochemical potential difference ($\Delta\mu$) of an ion X^+ across a membrane separating compartments A and B can be calculated as follows:
$\Delta\mu\ (X^+) = (RT\ln[X^+]_A/[X^+]_B + zF\ (E_A - E_B))$
(Where $\Delta\mu$ is the electrochemical potential difference between A and B; R is the ideal gas constant; T is absolute temperature; $[X^+]_{A/B}$ is the concentration of X^+ in A and B; z is valency; F is Faraday's number; $E_A - E_B$ is the electric potential difference across the membrane.)

When no net movement of X^+ across a membrane occurs it is at equilibrium and the electropotential difference ($\Delta\mu$) for X^+ is zero, therefore:

$$RT\ln \frac{[X^+]_A}{[X^+]_B} + zF(E_A - E_B) = 0$$

Solving for $E_A - E_B$ gives:

$$E_A - E_B = -\frac{RT}{zF}\ln\frac{[X^+]_A}{[X^+]_B} = \frac{RT}{zF}\ln\frac{[X^+]_B}{[X^+]_A}$$

A convenient form of the equation is obtained by converting to a form that involves $\log_{10}$ ($\ln y = 2.303 \log y$). At 29°C the quantity 2.303 RT/F is equal to 60 mV:

$$E_A - E_B = \frac{60\text{mV}}{z}\log\frac{[X^+]_B}{[X^+]_A}$$

Fig. 3.15 Derivation of the Nernst equation. $E_A - E_B$, the electric potential difference across the membrane; R, ideal gas constant; T, absolute temperature; $[X^+]_{A/B}$, concentration of X^+ in A and B; z, valency; F, Faraday number.

as the electrochemical potential of the ion on side A minus that of the ion on side B (Fig. 3.14), if $\Delta\mu$ is:

- positive – ions move from A to B
- negative – ions move from B to A
- zero – there is no net movement of ions (the solution system is at equilibrium).

When a potential difference for an ion exists across a membrane there is a tendency for it to move down a chemical or electrical gradient. This potential energy can be harnessed in secondary active transport.

The Nernst equation

When a reaction is at equilibrium there is no net movement of ions across the cell membrane, i.e. the concentration gradient and the electrical gradient are balanced. In this situation, the electrochemical potential difference equation above can be rearranged to give the Nernst equation (Fig. 3.15):

$$E_A - E_B = (60\text{ mV}/z)\left(\log\left([X^+]_B/[X^+]_A\right)\right)$$

> **HINTS AND TIPS**
>
> When moving down a chemical gradient, an ion is moving from an area of high concentration to low concentration. Ion movement down an electrical gradient is dictated by charge. An ion crossing a membrane down an electrochemical gradient is responding to both chemical and electrical gradients.

The Nernst equation can be used to calculate:

- the electrical potential difference that must exist between two chambers for an ion to be in equilibrium across the membrane

- the direction an ion will flow when the reaction is not in equilibrium, given an experimentally derived electrical potential difference.

The electrochemical potential difference when an ion is at equilibrium is called the equilibrium potential, e.g. $E_{Cl} = -70$ mV.

Gibbs–Donnan equilibrium
The Gibbs–Donnan equilibrium describes the electrochemical equilibrium that develops when two solutions are separated by a membrane that is impermeable to at least one of the ionic species present.

In an experimental system consisting of two compartments containing equimolar solutions of KCl (compartment B) and KY (compartment A), separated by a membrane permeable to K^+ and Cl^-, but impermeable to the Y^- anions, the permeant ions will redistribute (Fig. 3.16A).

- Cl^- moves down its concentration gradient from B to A.
- Electroneutrality is preserved, because K^+ follows, moving from B to A.
- Y^- cannot diffuse across the membrane, and it is trapped in compartment A
- Therefore, K^+ remains trapped in compartment A and electroneutrality is preserved.

Electroneutrality is preserved when ions cross the membrane because the movement of Cl^- sets up a local electrical potential that draws K^+ across the membrane. The movement of ions continues until the reaction is at equilibrium (Fig. 3.16B) at which point the tendency for:

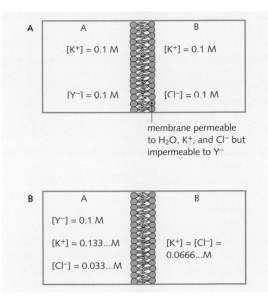

A

A	B
[K⁺] = 0.1 M	[K⁺] = 0.1 M
[Y⁻] = 0.1 M	[Cl⁻] = 0.1 M

membrane permeable
to H_2O, K^+, and Cl^- but
impermeable to Y^-

B

A	B
[Y⁻] = 0.1 M	
[K⁺] = 0.133...M	[K⁺] = [Cl⁻] = 0.0666...M
[Cl⁻] = 0.033...M	

Fig. 3.16 Gibbs–Donnan equilibrium. (A) Ion concentrations at the start of the experiment. (B) Ion concentrations on reaching Gibbs–Donnan. (Adapted from Berne et al., 1998.)

- Cl^- to move down the concentration gradient from B to A is offset by its tendency to move down the electrical gradient from A to B
- K^+ to move down its concentration gradient from A to B is offset by its tendency to move down the electrical gradient from B to A.

At equilibrium, $\Delta\mu\ K^+$ and $\Delta\mu\ Cl^-$ both equal zero. This is the basis for the Gibbs–Donnan equation, which states that the product of the concentrations of both permeant ions is the same in each compartment:

$$[K^+]_A[Cl^-]_A = [K^+]_B[Cl^-]_B$$

The Gibbs–Donnan equation holds for any univalent anion–cation pair in equilibrium between two chambers. A system in Gibbs–Donnan equilibrium has a number of important features.

- The compartment containing the impermeant ion contains more osmotically active ions (Fig. 3.16B).
- The compartment containing the impermeant anion has a negative electropotential. (whereas a compartment containing an impermeant cation would have a positive electropotential.)

Living cells resemble the experimental Gibbs–Donnan equilibrium above in a number of respects.

- They contain impermeant ions; proteins and nucleic acids.
- The membrane is permeable to K^+ and Cl^-, which are abundant.

However, there are significant differences.

- The cell is sensitive to osmotic gradients.

- The cell membrane is not entirely impermeable to positively charged ions, such as Na^+ and Ca^{2+}, which leak into the cell down an electrochemical gradient.
- If they accumulate, osmotic pressure is exerted and the cell swells.

Cell swelling is avoided by actively transporting such ions out of the cell. The Na^+/K^+ ATPase pumps three Na^+ ions out of the cell, while only two K^+ ions are pumped in.

Resting membrane potential

The membrane potential (E_m) of a cell is proportional to the concentration gradient of the dominant ions (Na^+, K^+ and Cl^-) and the membrane permeability to each one. If an ion is freely permeable across the cell membrane, it will tend to force E_m towards its own equilibrium potential. Resting excitable cells are most permeable to K^+, so E_m reflects the balance between K^+ leaking out of the cell down its concentration gradient and being pulled in down the electrical gradient (i.e. E_K). This movement is not accompanied by the extrusion of an anion because the membrane is only permeable to Cl^-, which has an opposing concentration gradient. Thus, E_m is proportional to the concentration of K^+ on each side of the membrane (Fig. 3.17).

Resting E_m (-70 mV) is not quite equal to E_K (-90 mV) because the membrane is slightly permeable to other ions, such as Na^+. Sodium influx makes the membrane potential slightly more positive because E_{Na} is positive (+60 mV), reflecting its tendency to move into the cell down its electrochemical gradient. However, Na^+ is not at equilibrium across the resting cell membrane because it is not freely permeable to this ion.

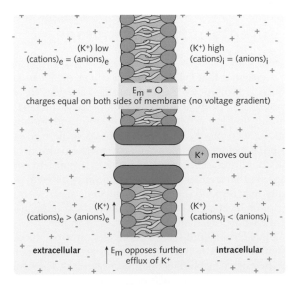

Fig. 3.17 Role of potassium in generating the cell membrane potential. E_m, membrane potential; e, extracellular; i, intracellular.

The action of the Na^+/K^+ ATPase, being electrogenic, contributes a small amount to the membrane potential directly. However, the majority of the membrane potential arises from the indirect action of the pump and reflects the movement of K^+ and Na^+ down the concentration gradients that it has established.

The excitable cell may manipulate its resting potential to control activity.

- Depolarization means the E_m becomes less negative, and the potential difference decreases. The nerve cell is more likely to fire.
- Hyperpolarization means that the potential difference increases in magnitude, by increasing the relative negative charge inside the cell. This nerve cell is less likely to fire.

(Action potentials are discussed in detail in *Crash Course: Nervous System*.)

HINTS AND TIPS

Membrane potential is highly sensitive to the concentration of K^+. An increase in the extracellular concentration of K^+ (hyperkalaemia) will partially depolarize excitable cells, bringing the resting potential closer to the threshold potential. Moreover, as K^+ efflux after an action potential is inhibited repolarization is impeded. Hypokalaemia (a decrease in the extracellular concentration of K^+) will tend to hyperpolarize cells, decreasing excitability. Both conditions affect cardiac cells, causing arrhythmias.

RECEPTORS

Concepts of transmembrane signalling

It is essential that cells in a multicellular organism are able to communicate to coordinate their activities. Signal transduction pathways regulate multiple cell activities including division, differentiation, migration and degranulation. Such processes enable responses to be made to external factors governing cell activity. The pathway begins at cell surface receptors and ends in the nucleus with proteins that regulate gene expression. Since different cell types may respond differently to the same signal at transcription level, these processes facilitate the coordination of the whole organism's response to a stimulus.

Only certain lipid-soluble molecules can cross the cell membrane directly (e.g. steroid hormones); others transfer their signal by binding cell surface receptors. The signal transduction pathways are formed by interacting proteins, which can amplify, dampen, or process signals before passing them downstream. Each cell may be confronted

with many different signals coming from its cellular neighbours, environment, substratum contact, and the presence of growth factors and hormones. The resulting signal pathways are integrated so that the cellular response is appropriate. When signal pathways malfunction, the cell may multiply uncontrollably, resulting in malignancy.

Important concepts include:

- Cell surface receptors are specific proteins that selectively bind a signalling molecule and convert this binding into intracellular signals, altering cell behaviour.
- Intracellular receptors are largely ligand-activated transcription factors which, when activated, migrate to the nucleus and bind to DNA, stimulating or suppressing gene transcription.
- The second-messenger system is a set of intracellular molecules activated by cell surface receptors and affecting cell function, producing a physiological response. Second messenger systems produce a signal cascade that amplifies the initial signal and facilitates various cellular responses, which vary between cell types.

The three mechanisms of cell signalling to surface receptors are (Fig. 3.18):

1. Endocrine
2. Paracrine
3. Autocrine.

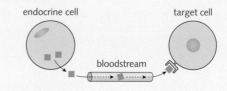

Endocrine signalling
hormone receptors are at a distance from endocrine cells

Paracrine signalling
local chemical mediators, which are metabolized more rapidly than hormones, act on local cells

Autocrine signalling
local mediators are produced by, and act on the same cell

Fig. 3.18 The three mechanisms of cell signalling. A single signalling molecule may fall into more than one of these categories depending on location of synthesis and release.

Types of receptor

The presence or absence of a specific receptor on a cell governs the responsiveness of that cell to signalling molecules. The majority of cell-surface receptor proteins belong to one of three main families:

1. Ionotropic
2. Metabotropic G protein-coupled
3. Enzyme-linked.

Ionotropic receptors

Ionotropic receptors or ligand-gated ion channels are similar to other ion channels, but contain a ligand-binding receptor site within their structure. They are composed of several subunits that, when activated by ligand binding, directly affect the activity of a cell by opening the ion channel. They are predominate in the nervous system where they mediate fast excitatory or inhibitory neurotransmission (e.g. nicotinic acetylcholine receptor; Fig. 3.19).

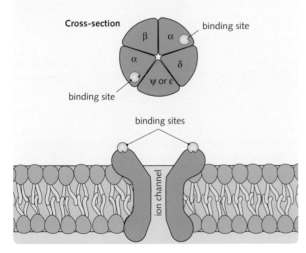

Fig. 3.19 Nicotinic acetylcholine receptor structure.

Metabotropic receptors (G-protein coupled)

Metabotropic receptors are membrane receptors that influence the activity of cells indirectly, with the transduction of an extracellular signal (ligand binding) to an intracellular one and the activation of second messenger molecules. This receptor class is defined by a common structure of a single polypeptide chain spanning the membrane seven times (Fig. 3.20). Metabotropic receptors activate GTP-binding proteins (G-proteins) (Fig. 3.21). G-proteins are heterotrimeric, composed of α-, β- and γ-subunits, and are bound to the cytoplasmic face of the plasma membrane. Receptor binding triggers a conformational change in the G-protein, with the dissociation of the α-subunit from the β- and γ-subunits, ultimately leading to GTP hydrolysis. Thus, they function like a binary switch (Fig. 3.20). Activated G-proteins subsequently activate intracellular second-messenger pathways, for example:

- cyclic AMP (cAMP) (Fig. 3.22)
- calcium (directly by opening plasma membrane calcium channels, or indirectly by the inositol lipid (IP_3) pathways)
- IP_3 pathways (Fig. 3.23).

Different G-proteins activate different pathways, for example:

- G_s increases cAMP
- G_i decreases cAMP
- G_q activates IP_3 pathways.

When GTP on the α-subunit is hydrolysed to GDP and Pi, it reassociates with the β- and γ-subunits and the binary switch is turned off. The β-adrenergic receptor is a metabotropic receptor.

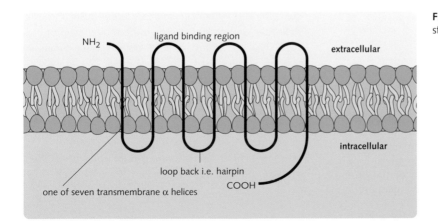

Fig. 3.20 Metabotropic receptor structure.

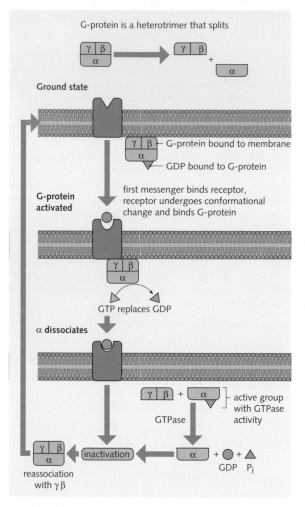

Fig. 3.21 G-protein activation. The inactive G-protein is a heterotrimer with GDP bound via its α-subunit. Interaction of the receptor with its ligand drives the exchange of GTP for GDP. This induces a conformational change in the α-subunit, resulting in its dissociation from both the receptor and the β-, γ-subunits. Dissociated subunits are free to interact with effectors that generate secondary messengers. Eventual hydrolysis of GTP by the α-subunit permits the regeneration of the inactive heterotrimer. GDP, guanosine diphosphate; GTP, guanosine triphosphate; Pi, inorganic phosphate.

Enzyme-linked receptors

Enzyme-linked receptors are single-pass transmembrane proteins, with extracellular and intracellular domains (Fig. 3.24). The intracellular domain either possesses intrinsic enzyme activity or associates directly with an enzyme, and is activated when the appropriate ligand binds to the external portion of the receptor. There are six classes of enzyme-linked receptor,

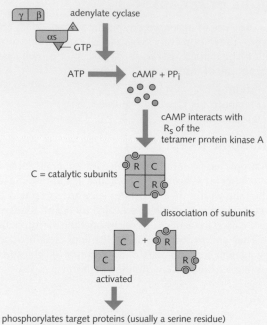

Fig. 3.22 Adenylate cyclase pathway. cAMP, cyclic adenosine monophosphate; ATP, adenosine triphosphate; C, catalytic subunits; GTP, guanosine triphosphate; PPi, pyrophosphate; R, regulatory subunits.

mediating a large number of ligands, including peptide hormones, growth factors and cytokines (Fig. 3.25):

1. receptor tyrosine kinases
2. tyrosine-kinase-associated receptors
3. receptor-like tyrosine phosphatases
4. receptor serine/threonine kinases
5. receptor guanylyl cyclases
6. histidine-kinase-associated receptors.

Clinical Note

Laron syndrome is an autosomal recessive disorder characterized by marked short stature. It results from molecular defects of the growth hormone receptor gene (type 1), or post-receptor defects in the signal transduction required to produce insulin-like growth factor-1 (IGF1) (type 2). Ordinarily, the interaction of growth hormone and the growth hormone receptor leads to activation of cytoplasmic tyrosine kinases. As mutations yielding Laron syndrome involve, or are downstream of, the growth hormone receptor, treatment with growth hormone does not increase the growth rate, and treatment with IGF1 is required.

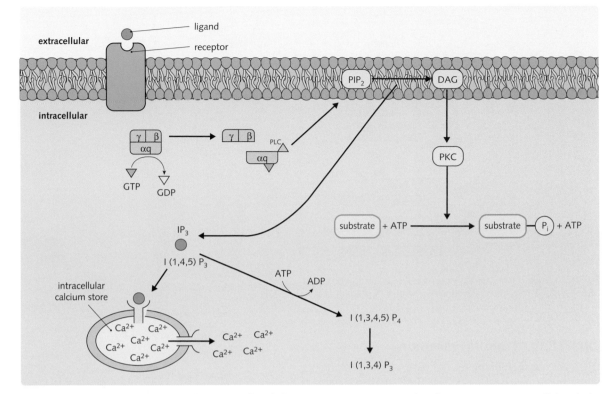

Fig. 3.23   Inositol phospholipid signalling. PLC, phospholipase C; GTP, guanosine triphosphate; ADP, guanosine diphosphate; PIP_2, phosphoinositol diphosphate; DAG, diacylglycerol; PKC, phosphokinase C; Pi, inorganic phosphate; IP_3, inositol triphosphate.

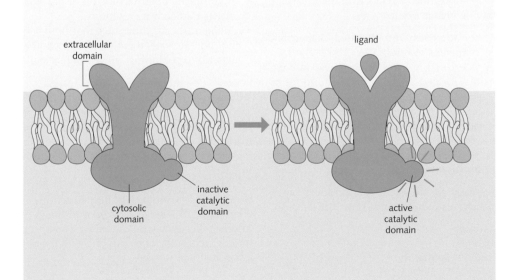

Fig. 3.24   Enzyme-linked receptor structure.

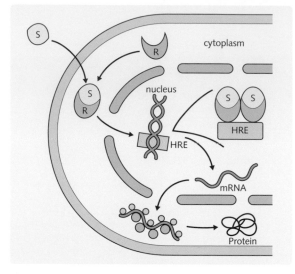

Fig. 3.25 Steroid receptor activations and cellular response. HRE, hormone-responsive element; R, receptor; S, steroid.

Intracellular (steroid) receptors

These are not membrane bound receptors, but soluble intracellular proteins (Fig. 3.26). There are two classes based on cellular localization. Class I (classical) steroid receptors are found cytoplasmically and are held in complexes with other proteins, such as heat shock proteins. They include receptors for the sex hormones and the steroid hormones of the adrenal cortex. Class II steroid receptors are located in the cell nucleus, and bind retinoid and thyroid hormones, and vitamin D.

When steroid hormones interact with their receptor, a characteristic sequence occurs.

- The receptor undergoes conformational changes and becomes competent to bind DNA.
- Activated receptors bind to hormone response elements (HREs) located in promoters of hormone responsive genes.
- The hormone-receptor complex functions as a transcription factor, activating or repressing associated gene transcription.
- The gene product may also regulate further gene transcription (the secondary response).

As such, members of the steroid hormone receptor family have three distinct domains (Fig. 3.27):

1. Ligand-binding
2. DNA-binding
3. Transcriptional regulatory.

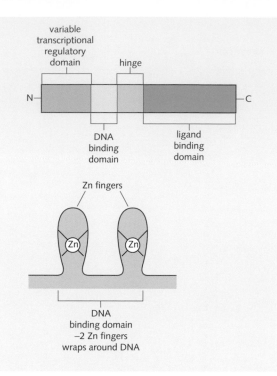

Fig. 3.26 Steroid receptor structure.

Receptors and drugs

Many drugs produce their pharmacological effect by acting on cell surface receptors. The effect depends upon whether the drug acts as an agonist or antagonist.

- Agonists are pharmacological or physiological molecules that activate receptors.
- Antagonists are molecules that bind receptors, but do not activate them. They block the receptor's ligand from binding, preventing its action.

Reversible antagonists, also called competitive antagonists, compete with the ligand for the receptor. Increasing the agonist's concentration can reduce this. Irreversible antagonists cannot be removed from the receptor; thus, they reduce the effective number of receptors, and agonist concentration has no impact.

The specificity of a drug reflects its ability to combine with one receptor type. The desired action of a drug is to combine with a specific receptor in the targeted tissue. Adverse effects may be caused by non-specific binding to other receptors, or by binding with the desired receptor in a different tissue.

Fig. 3.27 Features and examples of enzyme-linked receptors.

Class	Mechanism of activation	Examples
Receptor tyrosine kinases.	Activation of the receptor by ligand leads to oligomerization of the receptor and enzymatic activation of an intrinsic kinase resulting in the phosphorylation of tyrosine residues in intracellular signalling proteins.	The binding of epidermal growth factor (EGF) to the epidermal growth factor receptor leads to the proliferation of various cell types. The binding of insulin to the insulin receptor leads to the stimulation of carbohydrate utilization and protein synthesis
Tyrosine-kinase-associated receptors.	These receptors possess no catalytic activity of their own, and associate with cytoplasmic tyrosine kinases. Kinases associated with these receptors either belong to the Src family or the JAK family.	The binding of ligand, such as α interferon, induces the noncovalent association of the two separate cytokine receptor subunits and activation of the associated cytoplasmic kinase. Receptors include cytokine receptors e.g. TNF, antigen receptors e.g. CD4 and CD8, the growth hormone receptor and prolactin receptor.
Receptor-like tyrosine phosphatases.	Activation of the receptor leads to the dephosphorylation of tyrosine residues on cytosolic signalling proteins.	The leucocyte common antigen (CD45), when cross-linked by extracellular antibodies, becomes activated with the removal of phosphotyrosine residues from specific target proteins.
Receptor serine/threonine kinases.	Activation leads to the phosphorylation of serine/threonine residues on intracellular signalling proteins.	Following ligand binding, type I and type II TGFβ receptors form a heterotetrameric complex that activates intracellular second messengers of the Smad family. The Smads then translocate to the nucleus and initiate gene transcription. TGFβ superfamily members have important roles in a wide range of developmental processes including tissue differentiation, morphogenesis, proliferation, and migration. Loss of TGFβ activity has been implicated in tumorigenesis.
Receptor guanylyl cyclases.	Activation catalyses the cytosolic production of cGMP, which activates cGMP dependent kinase (PKG), which phosphorylates specific serine/threonine residues in target proteins.	Atrial natriuretic peptide (ANP) binding to the ANP receptor leads to transcriptional regulation of key genes and to counter the blood pressure-raising effects of the renin–angiotensin system.
Histidine-kinase-associated receptors.	A two stage reaction in which autophosphorylation of a histidine residue yields Pi, which is then transferred to a asp residue on a cytosolic signaling protein. Also known as two-component activation. Such systems are utilized for bacterial chemotaxis and are also seen in yeast and plants, but have not been not identified in man.	The Hog 1 osmoregulation pathway of *S. cerevisiae*.

By the end of this chapter you should be able to:
- Describe the basic structure of actin, microtubules and intermediate filaments.
- Understand the role of the cytoskeleton in the structure and function of cells and their specializations.
- Understand the structure and functions of lysosomes.
- Describe the basis of endocytosis, pinocytosis and phagocytosis.
- Understand the function of the lysosome and appreciate the different mechanisms leading to lysosomal storage diseases.
- Appreciate the function of the different types of cell–cell and cell–matrix junction.
- Understand the structure and function of adhesion molecules.
- Identify the components of the extracellular matrix and summarize their functions.

CYTOSKELETON AND CELL MOTILITY

Concepts

The cytoskeleton is a dynamic system of structural proteins that support the topography of the cell membrane, organizing the cytoplasmic components into defined areas. Its major functions are:

- determining cell shape
- providing mechanical strength
- organelle anchoring and polarity determination
- motility (and migration)
- anchoring of the cell to external structures
- metabolic functions
- separating duplicated chromatids and homologous chromosomes into separate cells at mitosis and meiosis respectively.

The major components of the cytoskeleton are:

- microfilaments – polymers of globular actin (G-actin)
- microtubules – polymers of α and β tubulin dimers
- intermediate filaments – polymers of a family of fibrous proteins that include the lamins and keratins
- in general, microfilaments have a structural role or are associated with cell movement; microtubules appear to be important in organelle organization and intracellular transport; and intermediate filaments provide the cell with mechanical strength.

Both actin and tubulin subunits have a pair of appropriately orientated, complementary binding sites that allow each subunit to bind to other monomers. In this way, long, helical structures are formed. As each subunit is asymmetrical, the resulting filament is polarized. The filaments are dynamic with one end (the plus end) being capable of rapid growth, and the other (the minus end) tending to lose subunits if not stabilized. Actin and tubulin are highly conserved throughout evolution, and they are found in all eukaryotic cells. However, by interacting with a range of accessory proteins, microfilaments and microtubules are able to perform a variety of distinct functions.

In contrast to actin and tubulin, the subunits of intermediate filaments are symmetrical, with a globular domain at each end. These fibrous monomers wind together to form the rope-like intermediate filaments.

Components of the cytoskeleton

Microfilaments

Actin microfilaments, with a diameter of 6.5 nm, form a layer just beneath the plasma membrane called the cortex. The individual globular subunits (G-actin) polymerize in a reversible process to form helical filaments (Fig. 4.1). This polymerization/depolymerization reaction is closely regulated by the cell; for example, extracellular signals may influence polymerization via G-protein coupled cell surface receptors, facilitating cell processes, such as chemotaxis in neutrophils.

The actin protein is encoded by a family of related genes. Mammals have at least 6 actin isoforms, which fit into three classes: α-actins, which are generally found in muscle, and β- and γ-isoforms, which are prominent in non-muscle cells. These isoforms may allow distinct protein interactions in accordance with the differing

Fig. 4.1 Structure of a microfilament (actin). Actin filaments consist of a tight helix of uniformly orientated actin molecules. The filament is extended as globular actin polymerizes at the plus end. Because of its appearance when complexed with myosin, the minus end is also referred to as the 'pointed end' and the plus end as the 'barbed end'.

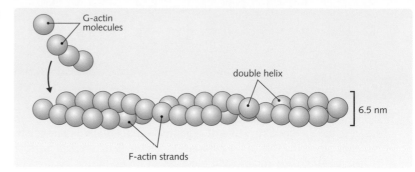

functions of the muscle and non-muscle forms. Actin has a contractile function in muscle cells. In non-muscle cells, actin:

- maintains structure of microvilli
- is a component of a specialized region of the cell cortex, the terminal web, which lies beneath microvilli and desmosomes
- facilitates movement of macrophages by gel–sol transitions of the actin network (gel phase – where the actin in the cytoskeleton is polymerized, to a sol phase – where it is soluble), mediated by actin-binding proteins
- facilitates movement of fibroblasts and nerve growth cones by controlled polymerization and rearrangement of actin filaments.

Various actin-binding proteins cause changes in the molecular forms of actin, and they can be classified into groups according to their function.

- Severing proteins such as gelsolin will cleave actin filaments in the presence of calcium ions. This property, when required, allows the cell to break up the cell cortex to facilitate processes such as phagocytosis.
- Linking proteins that bind actin strands together. Actin may be bound into tight arrays of parallel strands by 'bundling proteins', such as fimbrin and α-actinin. Alternatively, it may be arranged into a loose gel by 'gel-forming proteins', such as filamin, that bind crosswise intersections between strands.
- Myosin proteins are members of a protein family that move groups of oppositely orientated actin filaments past each other. This is the basis for contraction in muscle cells, but it is also important in non-muscle cells, where a transient assembly of actin and myosin produces the contractile ring that separates the cells in cell division. Other accessory proteins, such as troponin, affect actin and myosin interactions (see p. 42).
- Attachment proteins mediate linking of actin filaments to the plasma membrane – this group includes fodrin, talin and vinculin.

Intermediate filaments

Intermediate filaments (IFs) are 8–11 nm wide (Fig. 4.2). They are generally more stable than microfilaments and microtubules, and do not dissociate into monomers under physiological conditions. IFs are thought to be the major structural determinants in cells. There are various types of IFs, distinguishable by the protein from which they are made. These include:

- Keratins – there are many isoforms of keratin, which can be divided into soft 'cytokeratins' and hard 'hair keratins'. There are about 10 hard keratins, giving rise to nails and hair, and about 20 cytokeratins found more generally in epithelia lining internal body cavities.
- Lamins – these are found exclusively in the nucleus, which localize to two nuclear areas: the nuclear lamina and the nucleoplasmic veil. The nuclear lamina lines the inner surface of the nuclear envelope and is responsible for the disarrangement and reassembly of the nuclear envelope into vesicles during mitosis or meiosis.
- Neurofilaments – these are found in neuron axons. They are responsible for the radial growth of an axon and thus determine axonal diameter. They may account for the strength and rigidity of the axon.
- Glial fibrillary acidic protein (GFAP) – this is found in glial cells surrounding neurons.
- Vimentin – this is expressed in mesenchymal cells, such as fibroblasts, and in endothelial cells. These fibres often end at the nuclear membrane and desmosomes. They are closely associated with microtubules, and they form cages around lipid droplets in adipose tissue.
- Desmin – this is found predominantly in muscle cells. It forms an interconnecting network perpendicular to the long axis of the cell. Desmin fibres anchor and orientate the Z bands in myofibrils, thus generating the striated pattern.

Each cell type usually contains only one kind of intermediate filament. Rapidly growing cells and myelin-producing glial cells do not have intermediate filaments.

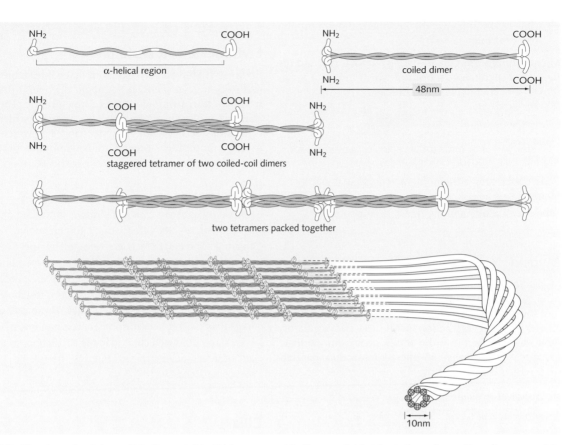

Fig. 4.2 Structure of an intermediate filament. Identical monomers bind in a parallel fashion to form dimers. Two dimers associate in antiparallel arrays to form tetramers, which wind together in groups of eight to produce the final rope-like intermediate filament. Since the association of the dimers in tetramers is antiparallel, intermediate filaments are not polarized. (Adapted from Norman and Lodwick, 1999.)

Microtubules

Microtubules are hollow tubules and are 25 nm wide (Fig. 4.3). They are polymers of tubulin dimers (α-β-dimer), and they extend from microtubule organizing centres, such as centrosomes, which stabilize the negative pole of the extending polymer. With the exception of mature erythrocytes, all cells have microtubules. They are particularly abundant in neurons, where they direct axon elongation.

There are several microtubule-associated proteins (MAPs), which have specific interactions with tubulin. Different microtubules associate with certain MAPs, e.g. tau in the nerve axon:

- some MAPs function as ATP-dependent molecular motors (e.g. dynein and kinesin). These motors may carry a cargo, such as an organelle or transport vesicle, along the microtubule to its designated location in the cell

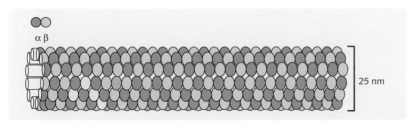

Fig. 4.3 Structure of a microtubule. There are normally 12–13 tubulin units per turn in the assembled microtubule. (Adapted with permission from *Molecular Biology of the Cell*, 3rd edn, by B Alberts et al, Garland Publishing, 1994. Reproduced by permission of Routledge, Inc., part of the Taylor & Francis Group.)

41

- some MAPs influence the polymerization of tubulin (e.g. centrioles).

Microtubules form cilia in the respiratory tract and the flagella of spermatozoa, both of which move via cycles of ATP-powered dynein arm linkage.

HINTS AND TIPS

The formation of microtubule spindles is essential for cell division. Nocodazole, taxol and vinblastine are antimitotic cancer chemotherapy drugs that interfere with the exchange of tubulin subunits between the microtubules and the free tubulin pool.

Myosin

Myosin is an actin accessory protein that functions as a molecular motor. It is composed of two heavy chains and four light chains (Fig. 4.4). The two essential light chains have ATPase action, while the two regulatory light chains determine the binding of calmodulin to myosin. Actin and myosin interact to produce contraction, which is regulated by:

- troponin in skeletal muscle
- calmodulin in non-muscle cells.

There are several isoforms of myosin, and muscle and non-muscle forms have slightly divergent amino acid sequences. (See *Crash Course: Musculoskeletal System* for further details.)

Examples of cytoskeletal function

Erythrocyte cytoskeleton

Erythrocytes have a very rigid, but malleable shape. The erythrocytic cytoskeleton is atypical, being present in only a thin strip below the cell membrane. The cell shape is indirectly maintained by spectrin, which directly links actin to ankyrin and band 4.1, which are, in turn, bound to integral transmembrane proteins (Fig. 4.5).

Clinical Note

Hereditary spherocytosis is a disorder of the red blood cell membrane, ultimately resulting in haemolytic anaemia. The key defects are cytoskeletal, most commonly spectrin deficiency. This causes membrane instability, the loss of erythrocyte surface area and abnormal cellular permeability to sodium, resulting in the production of rigid, spherical cells. These cells are fragile and susceptible to spontaneous haemolysis. They have a reduced lifespan in the circulation, as they are generally unable to pass through the splenic microcirculation.

Cilia

Cilia have a unique structure that aids the movement of cells or substances. The molecular biology of this movement is described on p. 48.

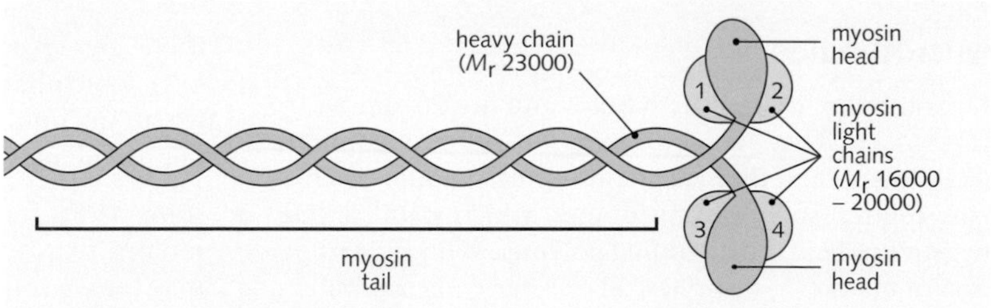

Fig. 4.4 Structure of myosin. Myosin II, which is the form found in muscle cells, is composed of two heavy and four light chains. The α-helices of the two heavy chains wrap around one another to produce a dimer. M_r, relative molecular mass. (Adapted from Stevens and Lowe, 1997.)

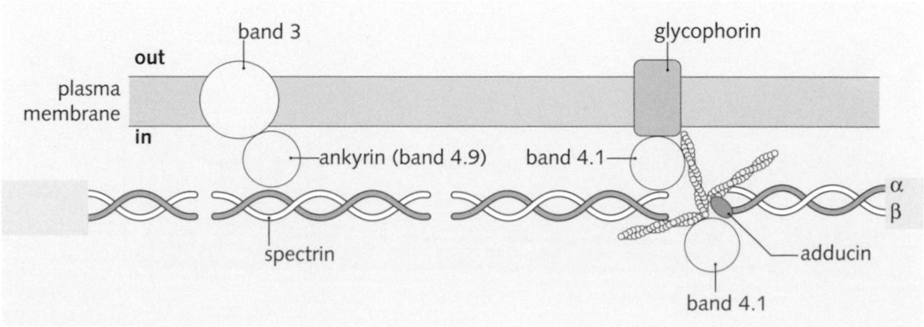

Fig. 4.5 The erythrocyte spectrin-based cytoskeleton. Spectrin is a dimer consisting of antiparallel α- and β-subunits. It is linked to the anion exchange protein (band 3) by ankyrin and to glycophorin by band 4.1, which also binds to actin and adducin (protein band numbers relate to migration in SDS-PAGE electrophoresis). (Adapted from Norman and Lodwick, 1999.)

Intestinal epithelium

Absorption is increased by microvillous projections, which increase intestinal surface area (Fig. 4.6).

Axonal transport

Kinesin and dynein transport materials along axons, each moving in a different direction. Organelle movement away from the cell body is driven by kinesin, which moves towards the plus end of the microtubule. Conversely, movement towards the cell body is driven by cytoplasmic dynein, which moves towards the minus end of the microtubule. Transport is normally at a rate of 25 mm/day (Fig. 4.7). Vesicles containing newly synthesized neurotransmitters are transmitted to the cell terminal by this means.

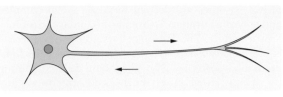

Fig. 4.7 Axoplasmic flow. Kinesin and dynein transport materials along axons.

Muscle contraction

In skeletal muscle, the arrangement of parallel actin and myosin into sarcomeres allows maximum efficiency of contraction. In smooth muscle, the contractile subunits resemble sarcomeres, but they are not as organized. (See *Crash Course: Muscles, Bone and Skin* for further details.)

Motility of phagocytes

Phagocyte motility is achieved by the projection of foot-like pseudopodia, which are associated with actin gel–sol transition at the tip, allowing the pseudopodia to advance.

Mitotic spindle

The spindle is a polar arrangement of microtubules across the equator of the cell. Chromosomes attach to the spindle via a kinetochore protein, at their centromeres. Separation of chromatids occurs as the microtubules contract, pulling them to separate poles (see Ch. 6).

LYSOSOMES

Definition

A lysosome is a membrane-bound organelle that contains acid hydrolases capable of breaking down macromolecules. Confinement of such enzymes in this organelle protects the rest of the cell from their potentially damaging effects.

Lysosomes have:

* diameters ranging from 50 nm to 1 μm
* a single membrane consisting of a phospholipid bilayer that undergoes selective fusion with other membranous organelles
* an ATP driven H^+ pump in the membrane, which acidifies the lysosomal matrix to pH 4.5–5.5, thus activating the hydrolases
* hydrolases in the inner matrix that are active at acid pH and break down carbohydrates, lipids, and proteins.

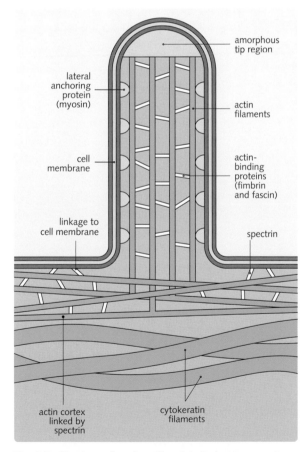

amorphous tip region

lateral anchoring protein (myosin)

actin filaments

cell membrane

actin-binding proteins (fimbrin and fascin)

linkage to cell membrane

spectrin

actin cortex linked by spectrin

cytokeratin filaments

Fig. 4.6 Structure of a microvillus. A helical arrangement of myosin molecules binds the actin bundle to the inner surface of the cell membrane. (Adapted from Stevens and Lowe, 1997.)

New lysosomes are derived from the Golgi complex and are called primary lysosomes. Secondary lysosomes are formed from the fusion of a lysosome with a vesicle containing substrate (Fig. 4.8). Most cells have hundreds of lysosomes, with phagocytic cells containing thousands. However, erythrocytes do not contain any lysosomes.

Functions of lysosomes

Lysosome functions are (Fig. 4.9):

- autophagy – digestion of material of intracellular origin (i.e. fuses with vacuoles from inside the cell)

- heterophagy – digestion of material of extracellular origin (i.e. fuses with vacuoles from outside the cell – pinocytic, endocytic, or phagocytic)
- biosynthesis – recycling the products of receptor-mediated endocytosis, which includes the receptor, its ligand and associated membrane.

Endocytosis is uptake of material into the cell, and it can be specific (receptor-mediated endocytosis) or non-specific (pinocytosis). Pinocytosis results in uptake of extracellular molecules at their extracellular concentrations. Phagocytosis is the internalization of membrane-bound particulate molecules by engulfment. It only occurs in specialized cells. The endocytic

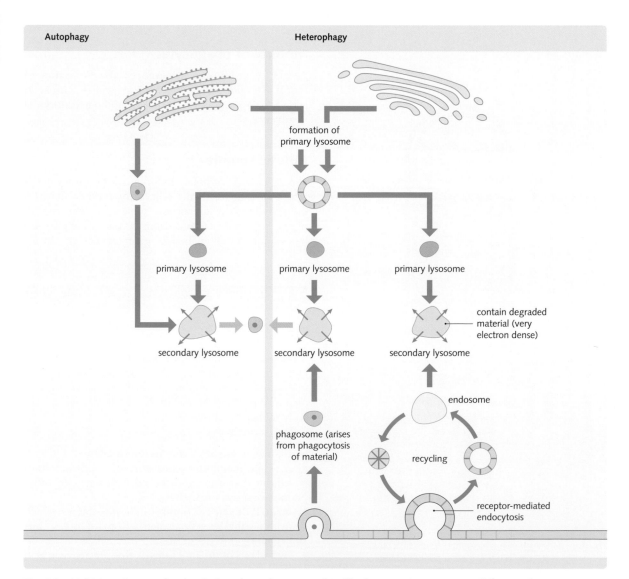

Fig. 4.8 Multiple pathways of endocytosis and membrane recycling. The lysosome is common to all these pathways.

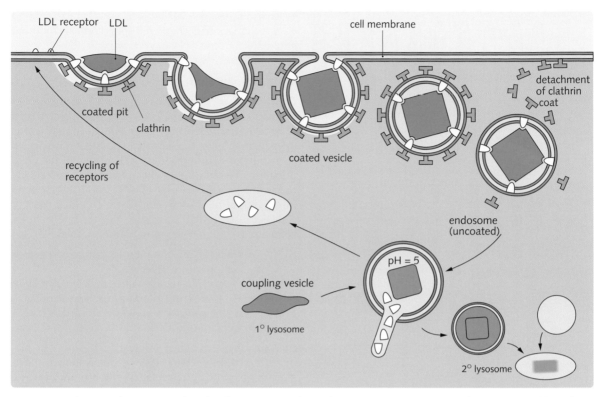

Fig. 4.9 Mechanism of receptor-mediated endocytosis. Low-density lipoprotein (LDL) receptors and their associated ligands are localized in clathrin-coated pits and are subsequently internalized in clathrin-coated vesicles. The coats are rapidly shed and uncoated vesicles fuse with endosomes. The LDL ligands dissociate from their receptors in the acid environment of the endosome and eventually end up in lysosomes. Meanwhile, the receptors are sequestered in a part of the endosome that is recycled back to the plasma membrane for reuse. (Adapted from Norman and Lodwick, 1999.)

vesicles that result from endocytosis fuse with primary lysosomes.

The products of enzymatic digestion are transported across the lysosomal membrane by specific receptors. Digestion is facilitated by lysosomal enzymes, of which over 60 exist, including:

- nucleases (e.g. acid RNase, acid DNase)
- glycosidases (e.g. β-glucuronidase, hyaluronidase)
- carbohydrate degradation enzymes (e.g. β-galactosidase, α-glucosidase)
- proteases (e.g. cathepsins, collagenase)
- phosphatases (e.g. acid phosphatase)
- sulphatases (e.g. aryl sulphatase)
- lipases.

Following synthesis in the rough endoplasmic reticulum (RER), lysosomal enzymes are modified by glycosylation in the RER lumen followed by covalent modification in the Golgi apparatus. Covalent modification includes phosphorylation of mannose groups to produce mannose- 6-phosphate groups, which act as recognition markers and direct the enzymes specifically to primary lysosomes.

Receptor-mediated endocytosis

Receptor-mediated endocytosis occurs when ligands that bind specific surface receptors are internalized in clathrin-coated pits (Fig. 4.10). In general, the ligand is degraded in the lysosome and its receptor is recycled to the cell surface. A variety of receptors and their ligands undergo receptor-mediated endocytosis (Fig. 4.11).

Lysosomal storage diseases

Lysosomal storage diseases (LSDs) are disorders of lysosomal function that result in macromolecules becoming trapped inside the lysosome. As such macromolecules build up, the lysosomes expand and the tissue enlarges, resulting in cellular dysfunction and pathological features. Such disorders may manifest as a result of:

- enzymatic processes – resulting from deficient or defective acid hydrolases, or absence of a crucial activator
- non-enzymatic processes – caused by transporter defects.

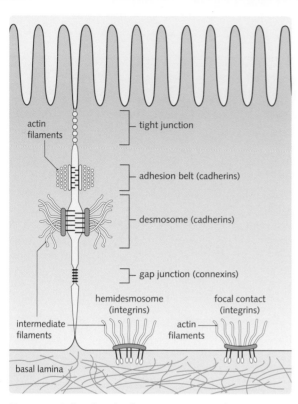

Fig. 4.10 Cell–cell and cell–matrix junctions. There are six distinct types of junctions in epithelial tissue. (Adapted from Norman and Lodwick, 1999.)

Fig. 4.11 Functions of receptor-mediated endocytosis.	
Molecules taken up	Function
Low-density lipoprotein (LDL)	Transports TAGs and cholesterol
Transferrin	Transports iron
Insulin	Affects cell metabolism
Fibrin	Removes injurious agents

There are some 40 LSDs, each of which is rare, but together they affect 1 in 4800 live births. They are commonly fatal, but can be treated with enzyme replacement therapy. All are single gene disorders and, with three exceptions, show autosomal recessive inheritance. Fabry and Hunter diseases are X-linked recessive disorders, while Danon disease shows X-linked dominant inheritance. Features leading to suspicion of a lysosomal storage disorder are:

- progressive neurological degeneration
- hepato(spleno)megaly
- skeletal dysplasia with or without short stature
- coarse facies
- eye changes (e.g. cherry red spot, corneal clouding)
- angiokeratoma.

Gaucher disease

Gaucher disease is the most common lysosomal storage disorder (incidence 1 in 25 000 live births), with a high incidence seen in Ashkenazi Jews, who have a carrier frequency of 1 in 60. A deficiency of β-glucosidase, which is encoded on chromosome 1q21, results in the accumulation of its substrate – glucocerebroside, principally in the phagocytic cells of the body, but also sometimes in the central nervous system. There are three types of Gaucher disease, defined by age of onset and brain involvement. Type 1 is the most common.

- Type I – adult type, non-neuronopathic. Lifespan is shortened, but not markedly.
- Type II – severe infantile, rare, neurological signs seen at 3 months, death usually by 2 years of age.
- Type III – 'juvenile' subacute, neuronopathic, variable presentation from childhood to 70 years of age.

Although the exact course of the disease cannot be predicted based on the genotype, key mutations in the β-glucosidase gene are associated with specific types of Gaucher disease, and can be useful in making clinical decisions. The substitution mutation N370S has a strong concordance with type I Gaucher disease, while the homoallelic L444P mutation (mutation at the identical site of both genes) correlates with severe, neuropathic forms, and the heteroallelic L444P mutation (mutation at different sites in both genes) correlates with milder forms of the disease, typically type III, although can also be found in type II. Tay–Sachs disease is another lysosomal storage disease that shows increased incidence in Ashkenazi Jews (1 in 25 carrier frequency). Carrier screening is available however the disease in invariably fatal by 3–4 years of age.

CELLULAR INTERACTION AND ADHESION

If groups of cells are to combine together to form organ structures, each cell has to be able not only to be held in its proper place, but also to communicate with its neighbours. Interactions between cells, and with the extracellular matrix (ECM), not only carry out a structural role, but may also facilitate cell–cell communication in several biological processes including migration, growth, immunological functioning, permeability, cell recognition, tissue repair, differentiation and embryogenesis.

Generally, cells bind via specific adhesion molecules to the ECM, providing elasticity and resistance to mechanical forces (i.e. as seen in connective tissue).

Epithelium has little ECM (only the basement membrane), so cell–cell interactions are adapted to bear tensile and compressive stresses, and show several types of cell–cell junctions.

Cell–cell junctions

Junctions are found between cells, and between cells and the ECM. There are three groups of cell junction (Fig. 4.12), which comprise six types. A junctional complex consists of a tight junction, an adhering junction and a desmosome.

Tight (occluding) junctions

All epithelia act as selectively permeable barriers, with tight junctions blocking diffusion of membrane proteins between apical (top) and basolateral (sides at the bottom) domains of the plasma membrane and sealing

neighbouring cells together so that water-soluble molecules cannot leak between cells (Fig. 4.13). Cell–cell contact at these junctions is mediated by the proteins occludin and claudins. The ability to restrict ion passage increases logarithmically with the number of occludin strands (e.g. small intestine tight junctions are 10 000 times more leaky than bladder tight junctions are). The degree of permeability offered by tight junctions is under physiological control and it is influenced by intracellular signals.

Anchoring junctions

Anchoring junctions are responsible for maintaining tissue integrity, and are most abundant in cells under stress (e.g. cardiac muscle). They link the cytoskeletons of adjoining cells to each other or to the ECM, and are made up of:

- intracellular attachment proteins
- transmembrane linker glycoproteins
- ECM or transmembrane linker glycoproteins on another cell (Fig. 4.14).

Anchoring junctions containing actin filament connections are called adherens junctions. They occur as streak-like attachments in non-epithelial cells and as continuous belts just below tight junctions in epithelial cells. These junctions attach the cytoskeletons (actin cell cortex) of adjacent cells together.

Desmosomes act as anchoring sites for intermediate filaments and, thus, provide tensile strength (Fig. 4.15). Cell–cell contact is mediated by desmogleins, which are a type of cadherin (see p. 49). Hemidesmosomes have a similar structure to desmosomes, but they link cellular intermediate filaments to the ECM (basement membrane) via integrin protein attachments (see Fig. 4.10).

Fig. 4.12 Types of cell junction.	
Group	**Members**
Occluding junctions	Tight junctions
Anchoring junctions	Actin filament attachment sites: (adherens junctions) cell–cell (e.g. adhesion belts) cell–matrix (e.g. focal contacts) Intermediate filament attachment sites: cell–cell (e.g. desmosomes) cell–matrix (e.g. hemidesmosomes)
Communicating junctions	Gap junctions Chemical synapses

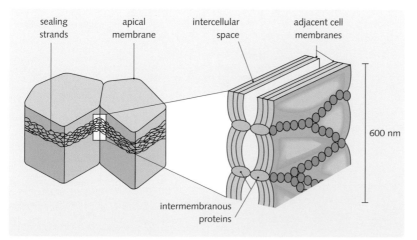

sealing strands · apical membrane · intercellular space · adjacent cell membranes · 600 nm · intermembranous proteins

Fig. 4.13 Structure of a tight junction. The tight junction forms a continuous band around the cell and it is, therefore, also called zonula occludens. The integral membrane protein occludin mediates cell–cell interaction. Each junction is made up of multiple pairs of this protein, one of each pair coming from each cell. (Adapted from Stevens and Lowe, 1997.)

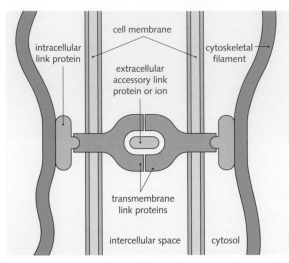

Fig. 4.14 General structure of an anchoring junction. Different (or multiple) link proteins and transmembrane proteins operate for the different classes of junction. (Adapted from Stevens and Lowe, 1997.)

Communicating (gap) junctions

In contrast to occluding and anchoring junctions, communicating junctions do not seal membranes together, nor do they restrict the passage of material between membranes. Gap junctions allow cells in a tissue to respond as an integrated unit. Inorganic ions carrying current and water-soluble molecules are able to pass directly from one cell to another through these structures, permitting electrical and metabolic cell coupling.

Clinical Note

The development of autoantibodies directed against the desmosomal proteins desmoglein-1 and desmoglein-3 leads to the pemphigus family of

'immunobullous' diseases. Autoimmune attack of these proteins leads to the separation of keratinocytes from each other (acantholysis), which float freely in the resultant blister. Autoantibodies to desmoglein-3 lead to pemphigus vulgaris, which, if poorly controlled, leads to secondary infections, disturbance of fluid and electrolyte balance, and can be fatal. Autoantibodies to desmoglein-1 results in the more superficial, and usually benign, pemphigus foliaceus.

One gap junction is composed of two connexons (or hemi-channels), which connect across the intercellular space (Fig. 4.16). Each connexon is formed from six connexins, and each connexin consists of four α-helices, various combinations of which combine to form gap junctions with different properties. Molecules of up to 1000 Da can pass through the pore, which is typically 1.5–2 nm in diameter. Some pores are gated, with opening related to a three-dimensional change, which is often mediated via extracellular signals. Several thousand connexons form a gap junction. Electrical coupling via gap junctions is important in:

- peristalsis
- synchrony of heart contractions
- coordination of ciliated epithelium.

Gap junctions also play a role in embryogenesis by allowing gradients of morphogens to form across blocks of cells.

Adhesion molecules

Adhesion molecules are cell surface ligands, usually glycoproteins, which mediate cell–cell adhesion. They have demonstrated roles in embryonic development, homeostasis, immune responses and malignant transformation.

Fig. 4.15 Structure of a desmosome. On the cytoplasmic surface of each interacting cell is a dense plaque composed of desmoplakin that is associated with attached intermediate filaments on one side and desmoglein (a type of cadherin) on the other. Cell–cell interaction is mediated by homophilic binding between adjacent desmoglein proteins. (Adapted from Stevens and Lowe, 1997.)

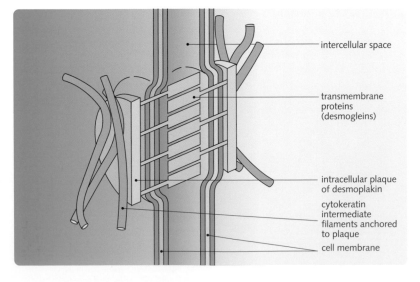

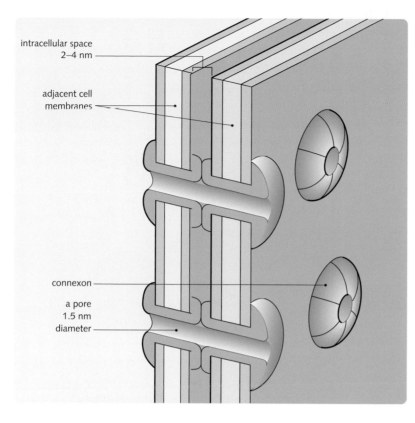

intracellular space
2–4 nm

adjacent cell
membranes

connexon

a pore
1.5 nm
diameter

Fig. 4.16 Structure of part of a gap junction. The junction consists of several hundred pores, which are aligned on adjacent cells. Each pore is composed of two connexons, one from each cell, which join across the intercellular gap to form a continuous aqueous channel. This channel facilitates electrical and chemical cellular coupling, since electrical currents and second messengers can pass freely through it. The cell can regulate permeability of gap junctions. (Adapted from Stevens and Lowe, 1997.)

Fig. 4.17 Families of adhesion molecules.

Family	Members	Ca²⁺/Mg²⁺ dependent	Cytoskeletal association	Associated cell function
Cadherins	E-CAD, N-CAD, P-CAD, desmosomal CAD	Yes	Actin filaments	Adhesion belt, desmosomes
Immunoglobulin (Ig) family	N-CAM, V-CAM, L1	Yes	Intermediate filaments (some members)	–
Selectins (blood and endothelial cells only)	P-selectin, E-selectin	No	–	Cell homing
Integrins	LFA-1 (β_2), MAC-1 (β_2)	Yes	Actin filaments, intermediate filaments	Focal contacts, hemidesmosome

There are four major cell adhesion molecule families (Fig. 4.17):

- cadherins
- immunoglobulin superfamily
- selectins
- integrins.

Cadherins

The cadherins are single-pass glycoproteins that mediate communication and adhesion (Fig. 4.18). Cadherins are generally involved in calcium-dependent homophilic interactions (i.e. the protein is both the ligand and the receptor). The N-terminal sequences, which contain

Fig. 4.18 (A) Structure of cadherin. It is composed of five extracellular domains, each 700–750 amino acid residues in length, and one intracellular domain. The intracellular portion is not present in T-CAD (T, truncated). (B) Cadherin exhibits homophilic binding, in which the molecule acts as both ligand and receptor.

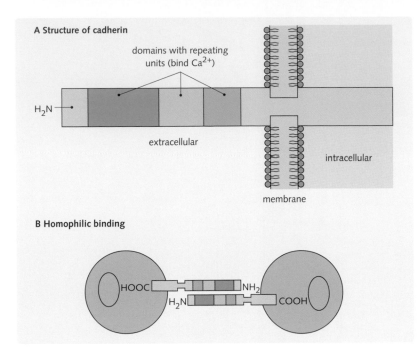

a conserved HAV (histidine, alanine and valine) motif, have been shown to be important in ligand binding and specificity. Cadherins are attached to the actin cytoskeleton by a class of linker proteins called the catenins.

Since cadherins are calcium dependent, changing the extracellular Ca^{2+} concentration alters their interactions. The three most widely expressed cadherins are:

- E-CAD (CDH1) – found in the epithelium and early nervous tissue
- P-CAD (CDH3) – found in the trophoblast (placenta) and epithelium
- N-CAD (CDH2) – found in nervous tissue and skeletal muscle.

Clinical Note

The ability of tumour cells for uncontrolled growth, migration, invasion and metastasis is often associated with disruption of cell–cell and cell–extracellular matrix junctions. The loss of cadherins can drive tumour invasion and malignancy by allowing easy disaggregation of cells. Reduced cell-surface expression of E-cadherin has been noted in many types of cancers, and arises either from somatic mutation in the E-cadherin gene, or as a secondary effect of mutations in other genes. Germline E-cadherin mutations yield a predisposition to gastric cancer.

Immunoglobulin (Ig) superfamily

These adhesion molecules are characterized by:

- antibody fold in each domain
- β-barrel structure
- two β-pleated sheets joined by cysteine–cysteine disulphide bonds, which are 60–80 residues apart
- loop regions without β-structure (variable expressed regions).

There is Ca^{2+} independent adhesion, and the Ig proteins have homophilic and heterophilic (protein is either a ligand or a receptor) binding sites. Ig members are involved in:

- adhesion
- signal transduction
- axonal growth and fasciculation (fasciculation means that axons grow along other axons by homophilic binding).

Important Ig family members include: neural cell adhesion molecule (NCAM), vascular adhesion molecule (VCAM), platelet endothelial (PECAM) and intercellular adhesion molecule (ICAM).

Selectins

Selectins are Ca^{2+}-dependent cell adhesion molecules which undergo heterophilic binding to carbohydrate

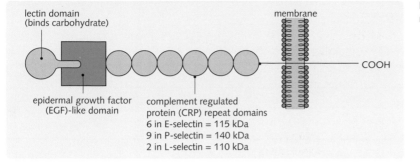

Fig. 4.19 Structure of a selectin molecule.

ligands (lectins) (Fig. 4.19), initiating leukocyte–endothelial interactions. There are three members of the selectin family:

1. L-selectin, which is expressed on leukocytes
2. E-selectin, which is expressed on activated endothelial cells
3. P-selectin, which is expressed on activated platelets and endothelial cells.

Expression of selectins is induced by local chemical mediators.

- E-selectin is activated by tumour necrosis factor (TNF), interleukin-1 (IL-1) and endotoxin.
- P-selectin is activated rapidly by histamine, thrombin, platelet activating factor and phorbol esters, and more slowly by TNF-α and IL-1.

The lectin domain recognizes specific oligosaccharides on the surface of neutrophils: the oligosaccharides Lewis X and sialyated Lewis X are recognized by P-selectin and E-selectin, respectively. These weak affinity interactions allow leukocytes to stick to the endothelial lining of blood vessels until integrins are activated. L-selectin is constitutive on the surface of polymorphonuclear neutrophils, monocytes and lymphocytes, facilitating the homing of these cells to lymph nodes and subendothelial capillaries.

HINTS AND TIPS

Do not be confused by the use of old nomenclature by books or tutors referring to the selectins. P-selectin is also called GMP-140, E-selectin is also called E-LAM.

Integrins

Integrins are integral plasma membrane proteins. They are the major receptors for binding to the ECM, and also have a role in signal transduction from the ECM to the cell. They consist of non-covalently associated heteroduplexes of α- and β-glycoproteins (Fig. 4.20). 18 α- and 8 β-glycoproteins have been identified, giving rise to at least 24 α–β pairs. Mammalian integrins form

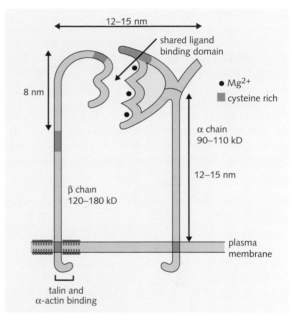

Fig. 4.20 Structure of integrin.

several subfamilies sharing common β-subunits that associate with different α-subunits. The integrins differ from other receptors in that they bind their ligand with low affinity, and they are present at high concentration. Interactions are heterophilic and Ca^{2+} or Mg^{2+} dependent, depending on the integrin (Fig. 4.21); β-2 and β-1 are important integrins.

β-1 integrins

These are found on most cells, forming dimers with at least 12 different α-subunits. The β-1 subunit is denoted as CD29, and β-1 integrins are often referred to as VLA (very late acting) molecules, as early *in vitro* experiments showed that they were expressed on T lymphocytes 2–4 weeks after stimulation. The most important is VLA4 (α-4-β-1), which binds VCAM-1 (a vascular adhesion molecule) and is important in homing lymphocytes to the endothelium at sites of inflammation.

Fig. 4.21 Integrins and transmembrane proteoglycans.

Family	Members	Ca^{2+}/Mg^{2+} dependent	Cytoskeletal association	Associated cell function
Integrins	Many	Yes	Actin filaments	Focal
Transmembrane proteoglycans	$\alpha_6\beta_4$	Yes	Intermediate	Hemidesmosomes
	Syndecans	No	Actin	None

β-2 integrins

These are exclusively expressed on leukocytes and, thus, are also referred to as the leukocyte function-associated antigen-1 (LFA1) family. They form dimers with at least four different types of α-subunit. LFA1 is also known as α-L-β-2. It mediates direct cell–cell interactions by binding intercellular adhesion molecules 1, 2 and 3 (ICAM-1, ICAM-2 and ICAM-3). MAC1 (macrophage antigen 1), exclusive to granulocytes and monocytes, is also a β-2 integrin and binds ICAM-1. Surface antigen ('cluster of differentiation') nomenclature is also used (e.g. CD18 is β-2).

Clinical Note

Failure to express the β-2 integrin (CD18) on leukocytes leads to the condition type 1 leukocyte adhesion deficiency (LAD1). As well as recognizing the ICAM molecules, the β-2 integrin is the receptor for complement (iC3b), and without it not only can leukocytes not migrate from the blood vessels to sites of infection, but they cannot recognize or bind complement either. This leads to an immunosuppressive phenotype and recurrent infection. Severe forms of the disease are associated with high mortality; patients typically succumb to bacterial infection within the first year of life.

Integrin binding

Integrins usually bind actin-based cytoskeleton inside the cell. Outside the cell, integrins can bind:

- ECM (e.g. fibronectin)
- cell-surface molecules (e.g. ICAM-1)
- soluble molecules (e.g. fibrinogen).

Integrins have recognition sites for ligands, e.g. Ig-like domains bind ICAMs and the tripeptide RGD (arginine, glycine, aspartic acid), which is a sequence commonly expressed in fibronectin and other ECM proteins. Cells can vary their binding properties by varying integrin affinities and specificities.

Integrins take part in signal transduction in cells (e.g. clustering of β-1-α-5 by fibronectin causes cytoplasmic alkalization).

Basement membrane

The basement membrane, comprising the basal lamina and the reticular lamina, is a sheet of ECM underlying epithelial and endothelial cells and surrounding adipocytes, Schwann cells and muscle cells. It acts to isolate these cells from the mesenchyme or connective tissue. It is composed of type IV collagen, heparan sulphate, proteoglycans, entactin and laminin. Functions of basement membrane are:

- cell adhesion
- to act as a porous filter in the kidney's glomeruli
- to inhibit the spread of neoplasia
- to regulate cell migration
- growth and wound healing
- differentiation.

Extracellular matrix

ECM is a hydrated polysaccharide gel containing a meshwork of glycoproteins. It is composed of:

- proteoglycans
- structural proteins – collagen, elastin
- fibrous adhesive proteins – laminin, fibronectin, tenascin.

ECM components are secreted by local fibroblasts in most tissues, but chondroblasts and osteoblasts are also involved in cartilage. ECM influences cell division, development, differentiation, migration, metabolism and shape. Connective tissues rely on the properties of the local ECM, which is:

- calcified in bone and teeth
- rope-like in tendon
- transparent in the cornea.

Proteoglycans

These form the hydrated polysaccharide gel that acts as a ground substance and allows diffusion of substances, such as nutrients and hormones, from the blood to

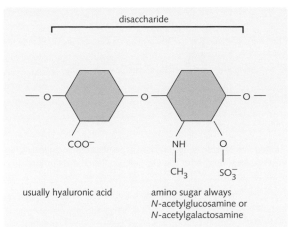

disaccharide

COO⁻   NH   O
                        | |
                       CH₃   SO₃⁻

usually hyaluronic acid   amino sugar always
                               N-acetylglucosamine or
                               N-acetylgalactosamine

Fig. 4.22 Glycosaminoglycan disaccharide subunit.

the tissue and vice versa. Glycosaminoglycans (GAGs) are unbranched polysaccharide chains of repeating disaccharide units (Fig. 4.22). Proteoglycans are formed in the Golgi apparatus where:

- the core protein is linked via a serine to a tetrasaccharide
- glycosyl transferases add sugar residues
- ordered sulphation and epimerization reactions (conversion of some amino acids from their natural L-isomer to the D-Isomer) occur.

The main types of GAGs are:

- hyaluronic acid – found as a lubricant in synovial fluid
- chondroitin sulphate – in cartilage
- dermatan sulphate
- heparin sulphate – an anticoagulant
- heparin
- keratin sulphate – in skin.

Chondroitin sulphate, dermatan sulphate and heparin sulphate are all between 500 and 50 000 Da.

GAGs are very hydrophilic and they have an extended coil structure, which takes up extensive space.

GAGs have a negative charge and attract cations, such as the osmotically active Na^+, bringing water into the matrix giving turgor pressure able to withstand forces of many hundreds of times atmospheric pressure. Proteoglycans function to:

- provide hydrated space
- bind secreted signalling molecules
- act as sieves to regulate molecular trafficking.

Proteoglycans and glycoproteins are compared in Figure 4.23.

Collagen

This fibrous protein has great tensile strength, and it is resistant to stretching (Fig. 4.24). It comprises 25% of the protein in mammals, and it is rich in proline (ring structure) and glycine (the smallest amino acid and occurring at almost every third residue so allowing the strands to fit together). Collagen synthesis is carried out in the ER and Golgi, as described in Chapter 2.

To date, over 20 different types of collagen have been described, encoded by a multigene family and divided into two main types: fibrillar and nonfibrillar. Types I, II, and III are fibrillar collagen and they are found in connective tissue (Fig. 4.25).

Fibrils are collagen aggregations of 10–300 nm in diameter and aggregate to form collagen fibres of a

Fig. 4.23 Comparison of proteoglycans and glycoproteins.	
Proteoglycans	**Glycoproteins**
Up to 95% carbohydrate	1–60% carbohydrate by weight
Unbranched carbohydrate	Branched carbohydrate
80 sugar residues	13 sugar residues
Larger than 3×10^5 Da	No larger than 3×10^5 Da

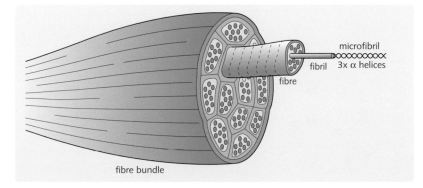

microfibril

fibril   3x α helices

fibre

fibre bundle

Fig. 4.24 Structure of collagen. Each individual collagen molecule is a left-handed helix. Three collagen molecules twist together to yield a microfibril. Microfibrils bundle together to give fibrils, which, in turn, bundle to give fibres.

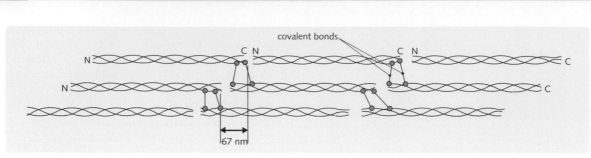

Fig. 4.25 Structure of fibrillar collagen. Collagen molecules are positioned side by side, staggered from adjacent molecules by one-quarter of their length. (Adapted with permission from *Molecular Cell Biology*, 2nd edn., by Darnell, Lodish and Baltimore, Scientific American Books, 1990.)

few millimetres in diameter. Organization is tissue specific; for example:

* 'wickerwork' pattern in skin resists multidirectional stress
* parallel layers in bone and cornea.

Type IV collagen is nonfibrillar, forming a sheet-like meshwork, and it is only found in the basal lamina (Fig. 4.26).

> **Clinical Note**
>
> Osteogenesis imperfecta (brittle bone disease) is a heterogeneous group of conditions characterized by spontaneous fractures, bone deformity and defective dentition. The substitution of a larger amino acid for glycine disrupts collagen triple helix formation and causes a severe, dominantly inherited form of the disease.

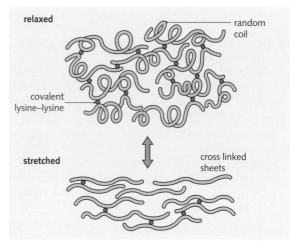

Fig. 4.27 Structure of stretched and relaxed elastin. (Adapted from Stevens and Lowe, 1997.)

Elastin

Elastin is found in places that need flexibility (e.g. skin, blood vessels and lungs). It is a highly glycosylated, hydrophobic protein (Fig. 4.27) that is rich in the non-hydroxylated forms of proline and glycine. One in seven amino acids is valine. The sheets are organized with the help of a microfibrillar glycoprotein, fibrillin,

which is secreted before elastin. Fibrillin deficiency results in Marfan syndrome.

Laminin

Laminin anchors cell surfaces to the basement membrane. It is a large heterotrimeric molecule, formed from

Fig. 4.26 Structure of type IV collagen, which assembles into multilayered sheets. (Adapted with permission from *Molecular Biology of the Cell*, 3rd edn, by B Alberts et al, Garland Publishing, 1994. Reproduced by permission of Routledge, Inc., part of the Taylor & Francis Group.)

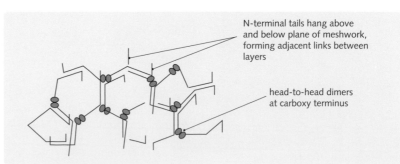

an α-, β- and γ-chain, held together in a 'cross shape' by disulphide bonds.

Fibronectin

This is an adhesive glycoprotein with binding sites for cells and matrix. It is a dimer of two subunits held together by disulfide bonds, which are folded into globular domains. Forms of fibronectin involved in wound healing and embryogenesis appear to promote cell proliferation and migration.

Tenascin

This protein, which can promote or inhibit cell adhesion and migration, is only produced by embryonic tissue and glial cells.

Role of the fibroblast

Fibroblasts are members of the connective tissue cell family. Members of the connective tissue family are all of common origin and interchangeable under appropriate conditions, other members being chondrocytes, osteocytes, adipocytes and smooth muscle cells. Connective tissue differentiation is controlled by cytokines, especially hormones and growth factors. Interchangeability allows them to support and repair most tissue types. Fibroblasts:

- secrete the fibrous proteins of the ECM in most tissues (except in cartilage and bone where they are produced by chondrocytes and osteocytes, respectively)
- are involved in the organization of ECM, enabling the configuration of ECM into tendons and other structures.

Macromolecules 5

Objectives

By the end of this chapter you should be able to:

- Explain the difference between an essential, semi-essential and a non-essential amino acid, and give named examples.
- Describe characteristics of amino-acid side chains and understand how they influence protein structure.
- Define what is meant by primary, secondary, tertiary and quaternary protein structure.
- Define the following: catalyst, enzyme, substrate, coenzyme and isoenzyme.
- Describe the interaction between the enzyme active site and its substrate.
- Define K_m and V_{max}.
- Describe a typical Michaelis–Menten graph and Lineweaver–Burk graph.
- Define the terms monosaccharide, disaccharide, polysaccharide, homopolysaccharide and heteropolysaccharide, and be able to give named examples of each.

AMINO ACIDS

Amino acids are the subunits of proteins, and they all have the same basic structure (Fig. 5.1):

- a central carbon atom (the a carbon)
- an amino (NH_2) group at the a carbon
- a carboxyl group (COOH)
- a side group (R).

There are 20 naturally occurring amino acids, which differ in their side group, the simplest being hydrogen (H) in the amino acid glycine. All amino acids, except glycine, have an asymmetrical α-carbon atom, giving rise to D or L stereoisomer forms; however, only the L form is found in humans (Fig. 5.2). Amino acids form proteins by joining together through peptide bonds that result from the condensation of the amino group of one amino acid with the carboxyl group of the next (Fig. 5.3). There may be a few to several thousand amino acid residues in a polypeptide, the sequence being determined by the base sequence in DNA. By convention, the free amino group is considered the start of a polypeptide sequence and the free carboxyl group its end.

Essential and non-essential amino acids

There are eight essential, two semi-essential and 10 non-essential amino acids in human biochemistry.

- Essential amino acids cannot be synthesized by the body, so they must be supplied by dietary protein.

These are tryptophan, lysine, methionine, phenylalanine, threonine, valine, leucine and isoleucine.

- Non-essential amino acids can be synthesized in the human body, so are not required in the diet. These are alanine, asparagine, aspartate, cysteine, glutamate, glutamine, glycine, proline, serine and tyrosine.

Semi-essential amino acids are not considered essential because humans can synthesize them *de novo*. However, the rate of biosynthesis does not increase to compensate for depletion or inadequate dietary supply. They become essential during times of growth, as the body cannot produce them in adequate amounts. These are histidine and arginine.

HINTS AND TIPS

Mnemonic for remembering the essential and semi-essential amino acids: *P*he, *V*al, *T*hr, *T*rp, *I*le, *M*et, *H*is, *A*rg, *L*eu and *L*ys (**PVT TIM HALL**).

Structure of amino acids

The amino acids can be classified by the nature of their R group or side chain (Fig. 5.4).

Properties of amino acids

The three-dimensional structure of a protein and, therefore, its behaviour, is determined by the characteristics and interactions of the side chains in its amino-acid sequence.

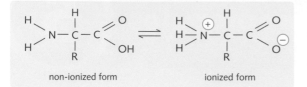

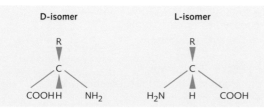

Fig. 5.1 Structure of an amino acid. R is the side group. In solution at pH 7.0 the amino acid exists in its ionized form. The charges on the amino and carboxyl groups disappear when they form peptide bonds.

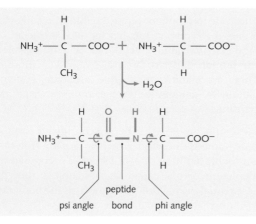

Fig. 5.2 D and L isomers of an amino acid. The stereoisomers are mirror images of one another.

Fig. 5.3 Condensation of amino acids into protein. The C–N bond is the peptide bond. This has a partial double bond character, so the atoms highlighted in bold remain in the same plane. Psi and phi angles are rotational angles about the indicated bonds. (Adapted from Stevens and Lowe, 1997.)

Size and structure

Larger, bulky side chains impede bending of the polypeptide chain, while ring structures prevent it forming the turns required to make α-helices ('steric hindrance'). Small or hydrophobic amino acids favour α-helical secondary structure.

Cross-linkages

Links between amino-acid residues occur through hydrogen bonds, disulphide bridges, hydrophobic bonds and ionic bonds, all of which act to stabilize the protein in its characteristic conformation.

- Hydrogen bonds occur between carbonyl (C=O) and imino (N–H) groups.
- Disulphide bridges are covalent bonds between thiol (-SH) groups of cysteine residues.
- Non-covalent hydrophobic bonds form between two hydrophobic residues.
- Electrovalent (ionic) bonds occur between a negative group of one amino-acid residue and a positive group of another amino-acid residue.

Solubility

In aqueous solution, colloidal proteins, such as cytoplasmic enzymes, are globular with polar R groups, which attract water, arranged on the outside and nonpolar groups arranged on the inside.

Covalent modification

Specific side groups can be modified by the addition of a chemical group in post-translational modification reactions (see p. 59), which alters the conformation of the protein, influencing its activity. For example, tyrosine residues are phosphorylated by tyrosine kinases, a property exploited in cell–cell signalling and in the regulation of enzyme activity (see Ch. 4).

Ionization properties of amino acids

Amino acids with non-polar R groups form amphions at neutral pH because the carboxyl group donates a hydrogen to the amino group, producing a molecule that has both a positively and a negatively charged group (see Fig. 5.1). This molecule is called a zwitterion, and its overall electrical charge is neutral. Amphions have the ability to act as both donors and acceptors of protons.

- At very low (acidic) pH both the amino group and the carboxyl group are protonated and the molecule (cation) has an overall positive charge.
- At very high (alkaline) pH both amino group and the carboxyl group are deprotonated and the molecule (anion) has an overall negative charge.
- Between these extremes the protonation of the amino and carboxyl groups and the overall charge of the molecule vary with pH.
- The pH at which the net charge on the molecule is neutral, such that the molecule would not move in an electric field, is called the 'isoelectric point'.

The Henderson–Hasselbalch equation

Depending on the pH, both the amino and the carboxyl groups of an amino acid may act as weak acids. A weak acid (HA) is one that only partially dissociates to its anion (A^-) and a proton (H^+). The value of the

Fig. 5.4 Classification of amino acids by side-group type. The individual side groups are highlighted by boxes.

Name	Symbol	Stereochemical formula	Side-group type
Aliphatic side chains			
Glycine	Gly (G)	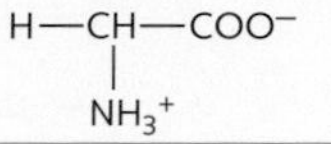 $H-CH-COO^-$ $\quad\;\; \vert$ $\quad NH_3^+$	Small
Alanine	Ala (A)	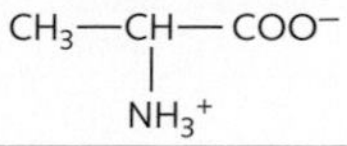 $CH_3-CH-COO^-$ $\qquad\;\; \vert$ $\qquad NH_3^+$	Hydrophobic +
Valine	Val (V)	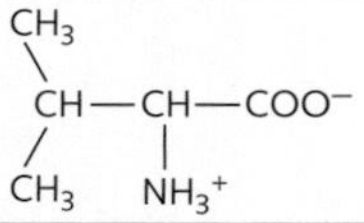 CH_3 $\;\;\backslash$ $\;\;\;CH-CH-COO^-$ $\;\;/\qquad\;\; \vert$ $CH_3\quad NH_3^+$	Hydrophobic ++
Leucine	Leu (L)	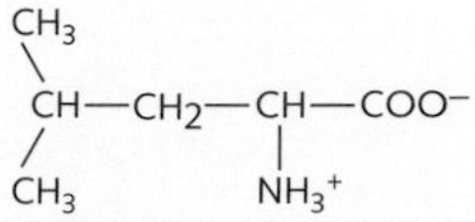 CH_3 $\;\;\backslash$ $\;\;\;CH-CH_2-CH-COO^-$ $\;\;/\qquad\qquad\;\; \vert$ $CH_3\qquad\quad NH_3^+$	Hydrophobic ++
Isoleucine	Ile (I)	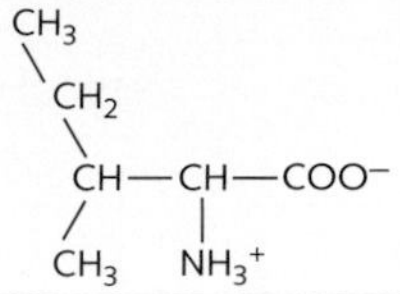 CH_3 $\;\;\backslash$ $\;\;\;CH_2$ $\quad\;\backslash$ $\quad\;\;CH-CH-COO^-$ $\;\;/\qquad\;\; \vert$ $CH_3\quad NH_3^+$	Hydrophobic +++
Aromatic rings			
Phenylalanine	Phe (F)	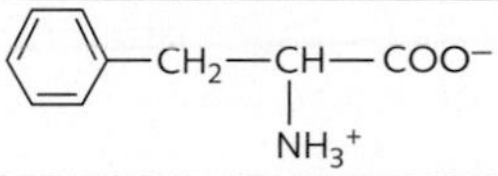 ⬡$-CH_2-CH-COO^-$ $\qquad\qquad\;\; \vert$ $\qquad\qquad NH_3^+$	Hydrophobic ++++
Tyrosine	Tyr (Y)	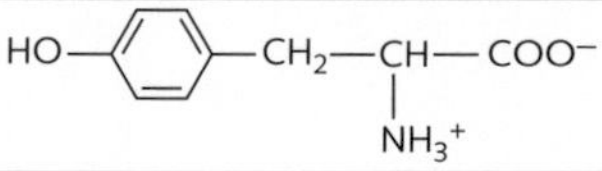 $HO-$⬡$-CH_2-CH-COO^-$ $\qquad\qquad\qquad \vert$ $\qquad\qquad\quad NH_3^+$	Hydrophobic (polar)
Tryptophan	Trp (W)	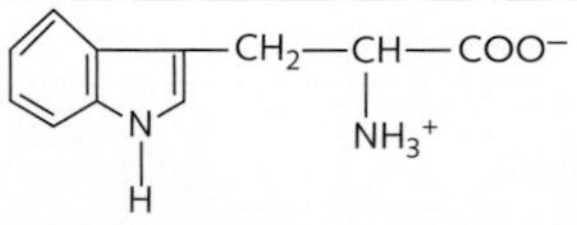 ⬡⬠$-CH_2-CH-COO^-$ $\qquad\qquad \vert$ $\qquad\quad NH_3^+$	Hydrophobic
Imino acids			
Proline	Pro (P)	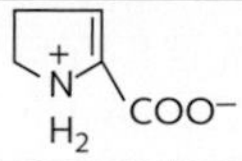 ⬠COO^-	Closed ring
Acidic groups or amides			
Aspartic acid	Asp (D)	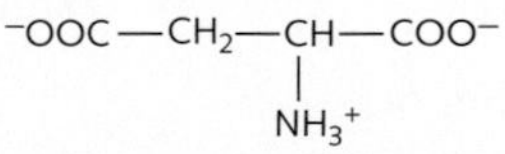 $^-OOC-CH_2-CH-COO^-$ $\qquad\qquad\;\; \vert$ $\qquad\qquad NH_3^+$	Weak acid, pK 4 negative charge
Asparagine	Asn (N)	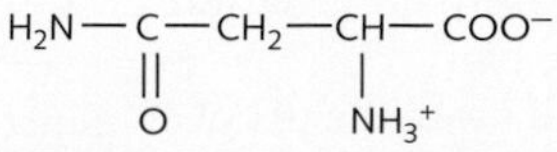 $H_2N-C-CH_2-CH-COO^-$ $\qquad \Vert\qquad\qquad \vert$ $\qquad O\qquad\quad NH_3^+$	Polar
Glutamic acid	Glu (E)	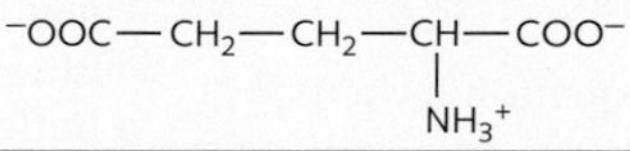 $^-OOC-CH_2-CH_2-CH-COO^-$ $\qquad\qquad\qquad\;\; \vert$ $\qquad\qquad\qquad NH_3^+$	pK 4 negative charge
Glutamine	Gln (Q)	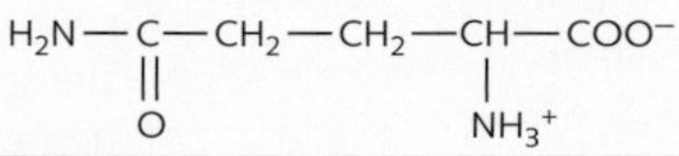 $H_2N-C-CH_2-CH_2-CH-COO^-$ $\qquad \Vert\qquad\qquad\qquad \vert$ $\qquad O\qquad\qquad\quad NH_3^+$	Polar

Continued

Fig. 5.4 Classification of amino acids by side-group type. The individual side groups are highlighted by boxes—cont'd

Name	Symbol	Stereochemical formula	Side-group type
Basic groups			
Arginine	Arg (R)	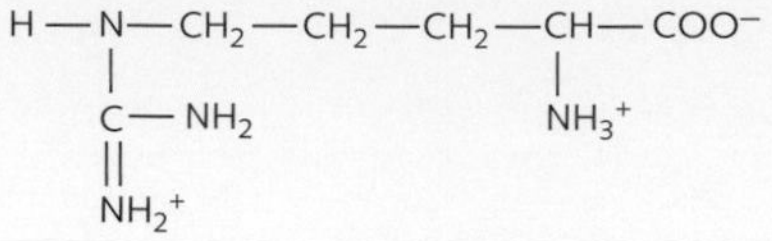 H—N—CH_2—CH_2—CH_2—CH—COO^- | | C—NH_2 NH_3^+ || NH_2^+	Weak base, pK 12 positive charge
Lysine	Lys (K)	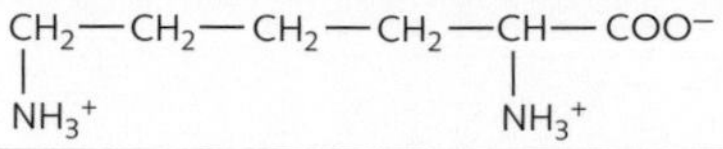 CH_2—CH_2—CH_2—CH_2—CH—COO^- | | NH_3^+ NH_3^+	pK 10 positive charge
Histidine	His (H)	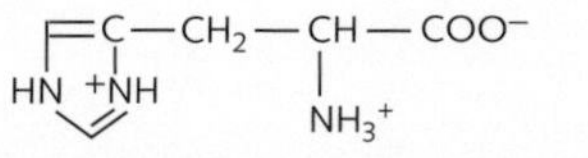 C—CH_2—CH—COO^- HN $^+$NH NH_3^+	pK 6 positive charge
Hydroxylic groups			
Serine	Ser (S)	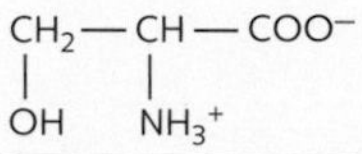 CH_2—CH—COO^- | | OH NH_3^+	Polar
Threonine	Thr (T)	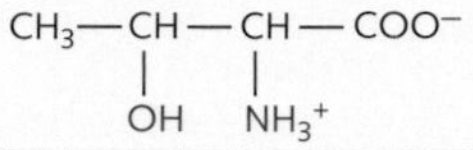 CH_3—CH—CH—COO^- | | OH NH_3^+	Polar
Sulphur groups			
Cysteine	Cys (C)	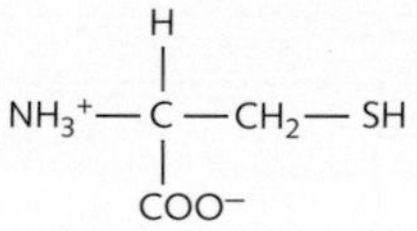 H | NH_3^+—C—CH_2—SH | COO^-	
Methionine	Met (M)	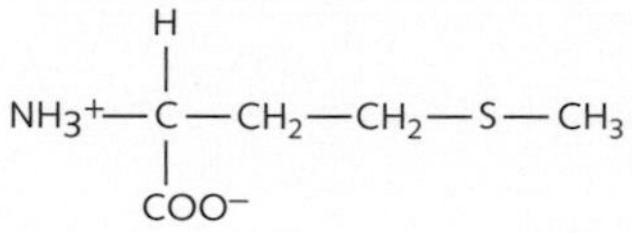 H | NH_3^+—C—CH_2—CH_2—S—CH_3 | COO^-	

dissociation constant (K_a) for a weak acid indicates its tendency to dissociate. Rearranging the formula for the dissociation constant gives the Henderson–Hasselbalch equation (Fig. 5.5):

$$pH = pK_a + \log[A^-]/[HA]$$

This equation permits the calculation of:

- the pH of a conjugate acid–base pair, given the pK_a and the molar ratio of the pair
- the value of pK_a for a weak acid given the pH of a solution of known molar ratio.

By the Henderson–Hasselbalch equation, when $[A^-]$ equals $[HA]$, pH equals pK_a, i.e. the pK_a of a weak acid is the pH at which it is half dissociated.

Amino acids as buffers

A buffer consists of a weak acid and its conjugate base. Buffers cause a solution to resist changes in pH when acid or base is added. In amino acids, the amino and the carboxyl groups may both act as buffers. At a pH of 9.8 the amino group of glycine functions as a buffer.

- At pH 9.8 glycine is in equilibrium between the anion and the zwitterions form.
- If protons are added they are removed from solution as they combine with the anion to produce the zwitterion.
- If alkali is added protons dissociate from the zwitterion to produce the anion and a water molecule.

Since at the pK_a of the amino group (9.8) there is an equal concentration of the weak acid (the zwitterion) and its conjugate base (the anion) it functions best as a buffer at this pH. A similar situation arises at pH 2.4 for the carbonyl group, though in this case the cation is the weak acid and the zwitterion its conjugate base. When glycine is titrated with an alkali:

- two pK_a values are observed at pH 2.4 (carboxyl group) and at pH 9.8 (amino group)
- the addition of alkali produces very little change in pH within one pH unit of each pK_a

$$HA \rightleftharpoons H^+ + A^-$$

weak   proton   conjugate
acid   base

The dissociation constant for a weak acid is:

$$K_a = \frac{[H^+][A^-]}{[HA]} = [H^+] - \frac{[A^-]}{[HA]}$$

Take the log of each of the terms in this equation.

$$\log K_a = \log [H^+] + \log \frac{[A^-]}{[HA]}$$

Rearrange thus:

$$-\log[H^+] = -\log K_a + \log \frac{[A^-]}{[HA]}$$

Substitute pH for $-\log [H^+]$ and pK_a for $-\log K_a$.

$$pH = pK_a + \log \frac{[A^-]}{[HA]}$$

Fig. 5.5 Derivation of the Henderson–Hasselbalch equation.

Fig. 5.6 Functions of proteins.

Protein function	Examples
Construction	Collagen in skin and bone
Contraction	Actin and myosin in muscle
Catalysis	Enzymes
Combat	Antibodies
Carriage	Haemoglobin in blood, transferrin carries iron
Communication	Peptide hormones, receptors, cytoplasmic kinases, major histocompatibility complex

Fig. 5.7 Proteins and ligands.

Protein	Ligand
Enzymes	Substrate
Myosin	Actin and other proteins
Antibodies	Antigen
Receptors	Hormones, neurotransmitters, counter-receptors

- the addition of alkali produces a large change in pH at the isoelectric point
- the isoelectric point lies midway between pK_{a1} and pK_{a2}.

Amino acids with ionizable side chains have a third pK_a corresponding to the pH range in which the proton on the side chain dissociates.

Non-protein amino acids

Several other amino acids are found in the body free or in combined states. These non-protein associated amino acids perform specialized functions. Biologically important non-protein amino acids include creatine, homocysteine, γ-aminobutyric acid and thyroxine.

PROTEINS

Functions of proteins

Proteins serve a variety of diverse functions in the human body (Fig. 5.6). Protein isoforms, related proteins encoded by different genes or generated by alternative RNA splicing of transcripts encoded by the same gene, often perform similar biological activities, but they differ in amino acid sequence, for example:

- myosin expressed in heart tissue
- myosin expressed in fast muscle fibres.

Protein conformation is defined by the sequence of its amino-acid residues, which is critical to its function (Fig. 5.7).

Organization of proteins

Primary structure

Primary structure is specified by the linear sequence of amino-acid residues (determined by the base sequence in DNA) linked via peptide bonds.

Secondary structure

Most proteins contain local regions of the polypeptide chain folding (α-helices and β-pleated sheets – see p. 62) resulting from hydrogen bonding interactions between peptide bonds (Fig. 5.8). They are favoured by:

- hydrogen bonding
- repulsion of side groups
- limited flexibility of the polypeptide chain.

Tertiary structure

Folding occurs to form the unique three-dimensional shape of a polypeptide chain. It is determined by interactions between side groups of the amino-acid residues, including disulphide bridges and electrostatic/hydrophobic interactions.

Quaternary structure

Two or more polypeptide chains (subunits) associate to form dimers, tetramers or oligomers (Fig. 5.9). The subunits are held together by the same types of bond that stabilize tertiary structure.

Fig. 5.8 Secondary structure of a protein. In both the α-helix and β-pleated sheet, regions of secondary structure are stabilized by hydrogen bonds (H bonds) between the C=O and N–H groups of the peptide bonds in the protein.

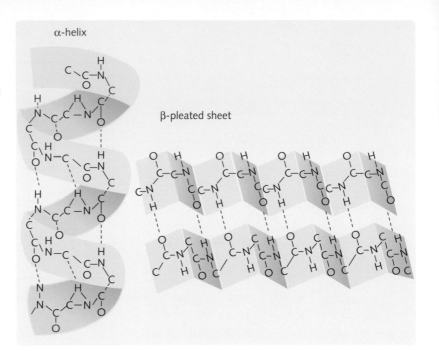

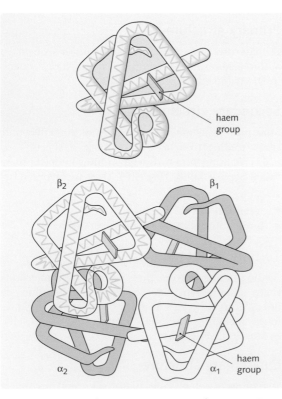

Fig. 5.9 Tertiary and quaternary structure of a protein. Four separate tertiary chains of haemoglobin are assembled into an oligomeric protein.

Forces that shape proteins

Peptide bond

The peptide bond is formed between two amino acids by condensation (Fig. 5.10). This is a strong covalent bond, which is resistant to heat, pH extremes and detergent, with a bond energy of 380 kJ/mol and length of 0.132 nm (1.32 Å). The peptide group is planar as it has a partial double-bond character. However, free rotation occurs around the other bonds, giving different phi and psi angles (Fig. 5.10) that allow considerable flexibility in the polypeptide chain. Consequences of the peptide bond are as follows.

- The polypeptide chain has considerable, though restricted, flexibility.
- The partial charge present at the oxygen and nitrogen of the bond enables attraction between two peptide bonds, forming a weak hydrogen bond with a bond energy of 5 kJ/mol.

Many biologically active proteins will spontaneously fold into one conformation that is stabilized by a variety of intramolecular bonds.

Hydrogen bonds

Hydrogen bonds occur between peptide bond atoms and polar side groups where a hydrogen atom is shared between two electronegative atoms; they are important

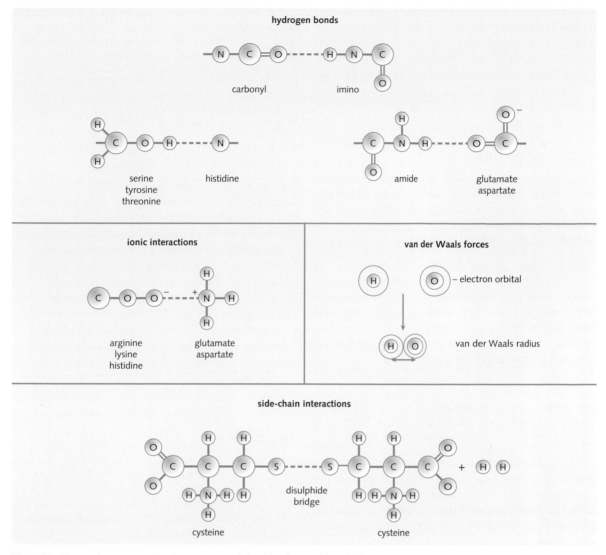

Fig. 5.10 Forces shaping proteins. Proteins are stabilized by chemical bonds that may depend on interactions between the carbonyl and imino groups in the polypeptide chain (e.g. some hydrogen bonds) or on specific side-chain interactions (e.g. disulphide bridges).

in forming secondary and tertiary structures. They have a bond energy of 20 kJ/mol and length of 0.3 nm (see Fig. 5.10).

Hydrophobic interactions

Hydrophobic residues (e.g. valine, alanine, leucine and phenylalanine) form interactions, rather than true bonds, where they cluster close together. Bond energy comes from the displacement of water.

Ionic interactions

Ionic interactions result from strong attractions between positive and negative atoms. These bonds are important in tertiary structure and have a bond energy of 335 kJ/mol and length of 0.25 nm (see Fig. 5.10).

van der Waals forces

van der Waals forces (dipole-induced dipole) are weak attractions between two atoms as the electron orbitals approach each other (see Fig. 5.10). The bond energy is very weak, being 0.8 kJ/mol, and the length is 0.35 nm. Collectively, these bonds 'add' to significant energy in the tertiary structure of large polypeptides.

Side-chain interactions

Side-chain interactions form bonds, the most important being between the thiol groups of two cysteine residues, forming a covalent bond of 210 kJ/mol called a disulphide bridge (see Fig. 5.10). Disulphide bridges are important in tertiary structures and in the secondary structure of elastin. This reaction is not favoured intracellularly, so disulphide bridges are normally found in exported proteins, for example:

- the digestive enzyme ribonuclease has four disulphide bonds
- the peptide hormone insulin has three disulphide bonds.

Protein folding

The correct 'folded shape' of a protein is determined by the amino acid sequence. Groups far away in the primary sequence may be brought close together in the final three-dimensional structure. Large proteins are composed of several domains linked by flexible regions of polypeptide. Domains have specific tertiary structure associated with a particular function, which may be conserved between proteins. For example, all known nicotinamide adenine dinucleotide (NAD) dependent dehydrogenase enzymes share a conserved NAD binding domain.

Structures within proteins

α-helix

The α-helix is a right-handed helix (D form) with a backbone of peptide linkages, from which side chains radiate outwards. Small or hydrophobic amino acid residues favour α-helix formation, so glycine or proline are usually found at the α-helix bends. The structure is stabilized by hydrogen bonds between every first and fourth amino acid, each hydrogen bond being relatively weak, but as all are parallel and 'intrachain' they provide reinforcement. The helix has:

- 0.15 nm rise
- 0.54 nm pitch
- 3.6 residues per turn.

The structure is a rigid rod-like cylinder that is very stable, with side groups pointing out and, therefore, free to interact with other α-helices (see Fig. 5.8). α-helices make up 90–100% of fibrous proteins and 10–60% of globular proteins. For example:

- α-keratin, a fibrous protein that is a component of skin, consists of long rod-like coils made from two identical α-helices wound around one another
- myoglobin, the globular protein that binds oxygen in muscle, has eight α-helical segments, and it is 75% α-helix in total.

β-pleated sheet

β-pleated sheets are extended chains formed from two or more pleated polypeptides joined by hydrogen bonds. In a parallel sheet the terminal amino acids

are at the same end whereas, in the more common antiparallel sheet, terminal amino acids are at opposite ends, so forming a more stable structure (see Fig. 5.8). The sheet is a rigid non-elastic 'platform', which is commonly found in fibrous proteins, for example:

- β-keratins in claws, scales, feathers and beaks are made of antiparallel strands
- silk is made of regular β-sheets.

Zinc fingers

Zinc fingers are a common motif in DNA-binding proteins such as transcription factors. The 'zinc finger domain' is a folded amino-acid projection surrounding a central zinc atom (Fig. 5.11). Zinc finger proteins recognize and bind to specific DNA regulatory sequences, influencing transcription.

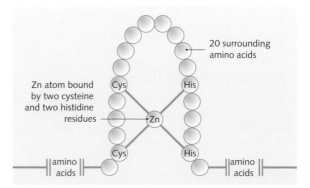

Fig. 5.11 Zinc finger domain. The bonds between the four amino acids and zinc stabilize a loop of polypeptide into a finger-like structure. The zinc atom may be coordinated by two Cys and two His residues, or by four Cys residues.

Collagen helix

Collagen is a fibrous protein that is a major component of connective tissue. It is composed of three polypeptide chains wound around one another and linked by hydrogen bonds (see p. 58).

> **Clinical Note**
>
> Tropocollagen is constructed from three left-handed helices twisted together into a right-handed helix. It is able to take on this unique shape as the side chain of every third amino acid is very close to the triple helix central axis. Glycine, having the smallest side chain of any amino acid, fulfils this requirement, and substitutions of glycine for bulkier amino acids, disrupting the architecture and collagen quality, have been implicated in osteogenesis imperfecta.

Stability of proteins

Denaturation is the loss of the three-dimensional structure of a protein due to breaking of structural bonds. It is usually associated with loss of biological activity (Fig. 5.12). Any treatment that disrupts chemical bonding (e.g. heat, pH extremes, detergents, oxidation, and physical effects such as shaking) may cause denaturation, thus most proteins only function within narrow environmental limits. If the denaturing conditions are not extreme (e.g. removal of the treatment), most proteins return to their active states when returned to optimal conditions; an example of this is lysozyme. Unlike insulin (Fig. 5.12), all the information required for lysozyme folding (tertiary structure) is present in its primary structure: thus, following denaturation and renaturation, fully functional lysozyme is recoverable.

Complex structures

Individual polypeptides may not be biologically active themselves, but they may serve as subunits in the formation of larger, active complexes. In addition, some proteins are dependent on their interaction with other, non-protein cofactors, such as prosthetic groups and coenzymes.

> **Clinical Note**
>
> Haemoglobin A is a tetramer composed of two α-subunits and two β-subunits each associated with a prosthetic haem group. Each haem group contains an iron atom, responsible for the binding of oxygen. Haemoglobin binds oxygen when pO_2 is high and releases it when it is low. Mutations in the haemoglobin genes result in an altered ability to carry and surrender oxygen to the tissues. The most common of these haemoglobinopathies are sickle-cell disease and β-thalassaemia.

ENZYMES AND BIOLOGICAL ENERGY

Properties of enzymes

Enzymes are biological catalysts. Without them, metabolic reactions would proceed too slowly for life. Enzymes have the following properties.

- They bind specific ligands (substrates) at active sites and catalyse their conversion to products.
- They greatly alter the speed of a reaction.
- They remain in the same chemical state at the end of the reaction as at the beginning, so can be reused.
- They catalyse the forward and the reverse reaction.
- They show great specificity. Some enzymes may only recognize one stereoisomer.

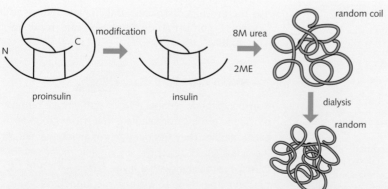

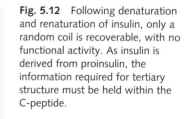

Fig. 5.12 Following denaturation and renaturation of insulin, only a random coil is recoverable, with no functional activity. As insulin is derived from proinsulin, the information required for tertiary structure must be held within the C-peptide.

An enzyme's name is normally the name of the substrate with the suffix -ase added, the substrate being the substance on which the enzyme acts. Isoenzymes catalyse the same reaction, but they have different primary structures and may work under different optimum conditions, for example:

- lactate dehydrogenase has heart and muscle isoforms
- creatine kinase has brain and muscle isoforms.

Mechanism of enzyme action

Catalysts and biological energy

In any chemical reaction, when a reactant is converted to a product, highly unstable intermediates are produced as chemical bonds are broken and reformed. These intermediates have higher free energy than the reactant, and so they are not favoured energetically (Fig. 5.13).

- The difference in free energy between the reactant and the unstable intermediate is the activation energy.
- Reactants must have in excess of the required activation energy.

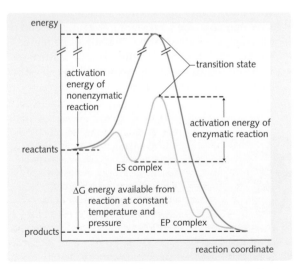

Fig. 5.13 Reaction profile for catalysed and non-catalysed reactions. Activation energy is less for the catalysed reaction. ES complex, enzyme–substrate complex; EP complex, enzyme–product complex. (Adapted from Baynes and Dominiczak, 1999.)

The energy level of individual reactants in a population follows a normal distribution. If activation energy is high, only a few molecules will have sufficient energy to react at any one time and the reaction will be slow.

A catalyst speeds up chemical reaction, but is itself unchanged by it. Catalysts bind transition molecules and stabilize them, so reducing the activation energy for the reaction. Therefore, in a catalysed reaction, more molecules will have the required activation energy and the reaction rate will increase. However, since they accelerate the reaction in both directions the position of the equilibrium is unchanged. A catalysed reaction has the following features relative to a non-catalysed reaction.

- The rate of the reaction is increased.
- The overall change in free energy between reactants and products is the same.
- The position of the equilibrium is the same.

Increasing the temperature of reactants increases their mean energy level. For a reaction catalysed by an inorganic catalyst, rate increases directly proportionally to temperature. However, enzyme-catalysed reactions show maximum activity at 37°C, but much reduced reaction rates at higher temperatures. This is because enzyme activity depends on the active site, and outside physiological parameters the bonds that maintain its structure are disrupted.

Active sites

The active site is the region to which substrate molecules bind. Enzyme specificity occurs because the shape of the active site is such that only substrates with a complementary structure can bind. This has been likened to the fitting of a key into its lock. However, rather than viewing the active site as a rigid structure the 'induced fit' model (Fig. 5.14) proposes that:

- the substrate binds to a substrate binding domain, which induces a conformational change at the active site
- the conformational change in the active site reveals functional groups
- as the product dissociates the enzyme returns to its original conformation and so can bind more substrate.

change, and are ultimately released as the reaction is completed. Alcohol dehydrogenase requires the presence of the coenzyme NAD.

Regulation of enzyme activity

The coordinated regulation of enzyme activity allows the organism to adapt to environmental change. In multi-step metabolic pathways the slowest enzyme determines the rate at which the final product is produced. The activity of such 'rate determining' enzymes is regulated so that metabolism is coordinated and energy is not wasted. Five mechanisms are involved:

1. Transcription from the enzyme coding gene may be regulated (see p. 86)
2. Enzymes may be irreversibly activated or deactivated by proteolytic cleavage
3. Enzymes may be reversibly activated and deactivated by covalent modifications, such as phosphorylation
4. Enzymes may be subject to allosteric modification. This occurs when the binding of a small molecule to a site distant from the active site alters the conformation of the enzyme and, therefore, its activity
5. The rate of degradation of the enzyme may be regulated.

Enzyme kinetics

Enzyme kinetics is the study of the rate of change of reactants and products. Enzyme assays use biosensors, oxygen electrodes, chromogenic substances and other methods to measure the progress of reactions.

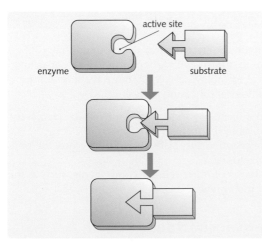

Fig. 5.14 The 'induced fit' hypothesis explains how the binding of the substrate to the enzyme changes the conformation of the active site.

Reaction rates

By plotting the amount of product formed against time, the initial velocity of the reaction can be estimated (Fig. 5.15). The initial velocity (V) equals the reaction rate. Reaction rate varies according to enzyme and substrate concentration. Increasing enzyme concentration increases the reaction linearly, so follows first-order kinetics (Fig. 5.15).

Increasing substrate concentration increases the reaction rate in an asymptotic, non-linear, fashion; as the enzyme becomes saturated the reaction rate reaches a limit – this is a Michaelis–Menten graph (Fig. 5.16). For simple enzymes the curve is a rectangular hyperbola. At low substrate concentrations the graph is linear (rate is proportional to [substrate]), so follows first-order kinetics. At high substrate concentrations a plateau is reached, so follows zero-order kinetics. The Michaelis–Menten equation relates the reaction rate to the substrate concentration:

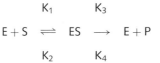

$$E + S \underset{K_2}{\overset{K_1}{\rightleftharpoons}} ES \overset{K_3}{\underset{K_4}{\longrightarrow}} E + P$$

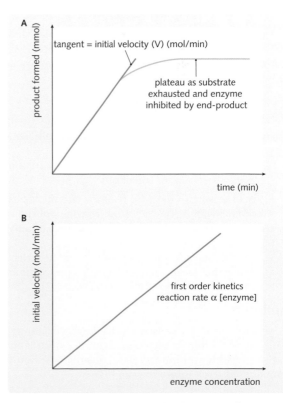

Fig. 5.15 (A) Calculation of initial velocity. (B) Effect of enzyme concentration on reaction rate.

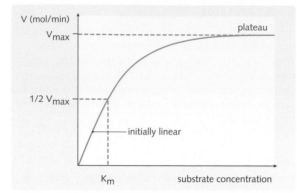

Fig. 5.16 Michaelis–Menten graph showing effect of increasing substrate concentration against reaction rate. K_m, Michaelis constant; V, initial velocity; V_{max}, maximum velocity.

K_1, K_2, etc., are individual rate constants. K_4 is insignificant so is ignored.

The Michaelis constant (K_m) is the rate of breakdown of the enzyme–substrate complex:

$$K_m = \frac{K_2 + K_3}{K_1}$$

The maximum velocity (V_{max}) is reached under saturating conditions when the substrate concentration is high:

$$V_{max} = K_3[ES]$$

where K_3 is rate of enzyme and product formation and [ES] is enzyme–substrate complex concentration.

The velocity (V) of the Michaelis–Menten reaction is therefore:

$$V = \frac{V_{max}[S]}{K_m + [S]}$$

where V_{max} and K_m are constants for each different enzyme.

For most enzymes, K_m is the substrate concentration at which the reaction rate is half of V_{max}.

- K_m is a measure of the affinity of the enzyme for its substrate, with a low K_m (low enzyme–substrate complex breakdown) corresponding to high affinity (tight binding) and vice versa.
- V_{max} and K_m are difficult to estimate from a Michaelis–Menten graph, so an alternative graph representation is used, the Lineweaver–Burk graph (Fig. 5.17).

Inhibitors

Enzyme inhibitors are substances that lower enzyme activity. False inhibitors include denaturing treatments and irreversible inhibitors (e.g. organophosphorus compounds). True inhibitors can be:

- competitive
- non-competitive
- allosteric.

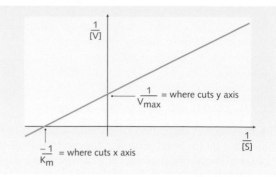

Fig. 5.17 Lineweaver–Burk plot. K_m, Michaelis constant; S, substrate; V, initial velocity; V_{max}, maximum velocity.

Competitive inhibitors resemble the substrate, and they compete for the active site (Fig. 5.18). The K_m is increased, so affinity of the enzyme is decreased; for example, azidothymidine (AZT) used to treat HIV infection resembles deoxythymidine, so it is a competitive inhibitor of HIV reverse transcriptase.

Non-competitive inhibitors (e.g. heavy metals such as lead) do not resemble the substrate, so they do not compete for the binding site. They bind to the enzyme and abolish its catalytic activity, although substrate may still bind (Fig. 5.18). V_{max} is decreased as there is less catalytically active enzyme, but K_m is unchanged as affinity of the enzyme is not altered.

Allosteric inhibitors do not bind to the active site; instead, they bind to a different allosteric site elsewhere on the enzyme. Binding causes a conformational change, which can alter V_{max} and/or K_m; for example, phosphofructokinase I (PFK I) is allosterically inhibited by high concentrations of ATP and by citrate (Fig. 5.19).

CARBOHYDRATES

Carbohydrates are chemical compounds that contain oxygen, hydrogen and carbon atoms, with the general chemical formula $C_n(H_2O)_n$. They have several important biological functions, including:

- metabolism to produce energy (ATP) by glycolysis
- storage in the form of glycogen
- conversion to fatty acids and triacylglycerol for long-term storage
- synthesis of other cellular components, such as the cell membrane.

Monosaccharides

Monosaccharides are the subunits of carbohydrates. They can be distinguished from each other on one of four levels:

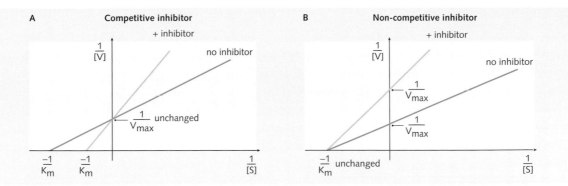

Fig. 5.18 (A) Competitive inhibition of enzyme. The inhibitor increases K_m. V_{max} is unchanged. (B) Non-competitive inhibition of enzyme. The inhibitor decreases V_{max}. K_m remains unchanged. K_m, Michaelis constant; S, substrate; V, initial velocity; V_{max}, maximum velocity.

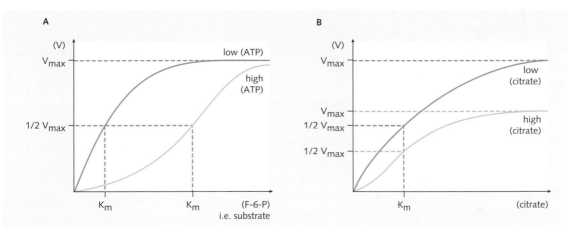

Fig. 5.19 Allosteric inhibition of phosphofructokinase 1. (A) ATP increases K_m (decreases affinity), but it does not alter V_{max}. (B) Citrate decreases V_{max}, but it does not alter K_m. K_m, Michaelis constant; S, substrate; V, initial velocity; V_{max}, maximum velocity.

1. Chemical nature of the carbonyl group (aldehyde to give an aldose or ketone to give a ketose)
2. Number of carbon atoms
3. Stereochemistry
4. Linear or cyclic nature.

Monosaccharides are polar due to their high proportion of hydroxyl groups, and as such are highly water soluble. With the exception of dihydroxyacetone, all carbohydrates contain at least one asymmetrical (chiral) carbon and, therefore, D and L stereoisomers exist. Unlike amino acids, in humans most carbohydrates exist in the D form, a notable exception being L-ascorbic acid.

Cyclic structures

Monosaccharides with five or six carbon atoms can form cyclic structures. In solution, both the cyclic and the straight chain forms co-exist, with the cyclic form predominating (Fig. 5.20). The aldehyde or ketone group of the straight chain structure may react with a hydroxyl group on a different carbon atom, forming a heterocyclic ring. Rings with five carbon atoms are called 'furanose' rings and those with six, 'pyranose' rings. Each ring structure has one more optically active carbon than the straight-chain form, and so has both an α- and a β-enatomer, which interconvert in equilibrium (Fig. 5.20).

Disaccharides and oligosaccharides

Disaccharides are formed by the joining of two monosaccharides with the elimination of water. This condensation reaction yields a glycosidic bond between the two subunits (Fig. 5.21), which exists in two forms, the α-form and the β-form; α-glycosidic bonds involve C1 of the α-anomer, while β-glycosidic bonds involve C1 of the β-anomer. Examples of important disaccharides are given in Figure 5.21.

Fig. 5.20 The relationship between straight chain and cyclical monosaccharides.

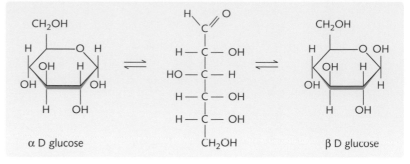

α D glucose   β D glucose

Fig. 5.21 Common disaccharides.

Disaccharide	Monomers	Linkage	Structure
sucrose	glucose + fructose	alpha 1–2	glucose / fructose
lactose	galactose + glucose	beta 1–4	galactose / glucose
maltose	glucose + glucose	alpha 1–4	glucose / glucose
isomaltose	glucose + glucose	alpha 1–6	glucose / glucose
cellobiose	glucose + glucose		glucose / glucose

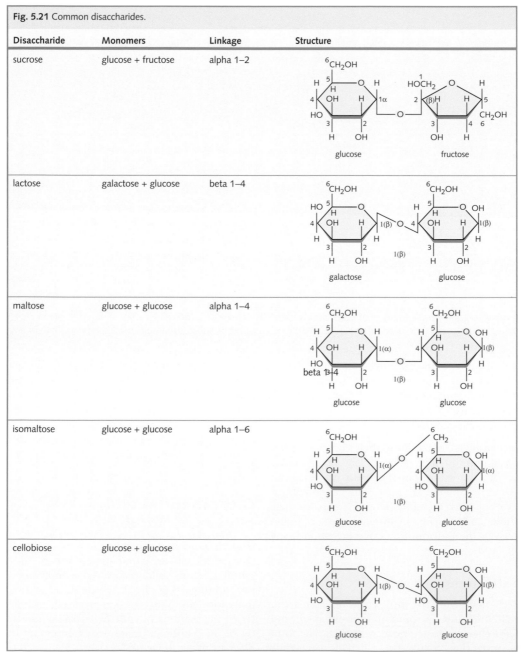

Clinical Note

Lactose is a disaccharide broken down by lactase, an enzyme secreted into the lumen of the small intestine. Most mammals stop producing lactase after infancy; however, a large number of the human population have an autosomal dominant mutation that enables them to continue. Those who do not are said to be lactose intolerant and, as lactose remains in the small intestine, it becomes subject to degradation and fermentation by bacteria, causing abdominal pain and distension and chronic diarrhoea.

Oligosaccharides are short chains of condensed monosaccharides, typically three to seven units long. They have a number of important biological functions, often covalently attached to proteins or to membrane lipids.

Polysaccharides

Polysaccharides are polymers of numerous monosaccharide units, as many as 2500, to yield a polymer with the general formula $C_n(H_2O)_{n-1}$. They can be classified according to whether they are:

- linear or branched
- the nature of glycosidic bond
- homopolysaccharide (repeating units of the same monosaccharide) or heteropolysaccharides (repeating units of different monosaccharides).

The main storage polysaccharide in animals is glycogen, a highly branched homopolysaccharide of repeating glucose units. The majority of the subunits are joined by α-1–4-glycosidic bonds. However, at every 10 residues or so falls a branch point, joined to the main molecule by an α-1–6-glycosidic bond. Cellulose, the main structural material of plants, is also a homopolysaccharide of repeating glucose units, but joined by β-1–4

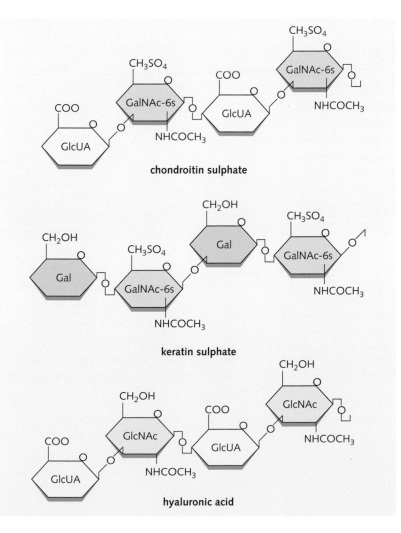

Fig. 5.22 Structures of common glycosaminoglycan units.

glycosidic bonds. Humans do not have a cellulase capable of digesting such links.

The most abundant heteropolysaccharides in the body are the glycosaminoglycans (GAGs). They are long unbranched molecules consisting of a repeating disaccharide unit of an amino sugar (an N-acetyl-hexosamine) and a sugar or sugar acid (a hexose or a hexuronic acid) (Fig. 5.22).

Clinical Note

Heparin is a glycosaminoglycan. The release of heparin from mast cell granules in response to injury leads to its entry into blood serum and the inhibition of blood clotting. Heparin acts by binding to antithrombin III, exposing its active site and allowing it to inactivate key proteases involved in blood clotting. This property has been manipulated for therapeutic gain, with commercially produced heparin being utilized as an antithrombotic agent for use in a range of venous and arterial clotting disorders.

Sugar derivatives

There are various sugar dervatives (Fig. 5.23), including:

- sugar acids: the C1 aldehyde group or the hydroxyl on the terminal carbon is oxidized to a carboxylic acid – e.g. ascorbic acid and glucuronic acid
- sugar alcohols: formed by the reduction of the carbonyl group of a sugar to a hydroxyl group – e.g. sorbitol, mannitol and ribitol
- amino sugars: one hydroxyl group is substituted for an amino group – e.g. glucosamine and galactosamine

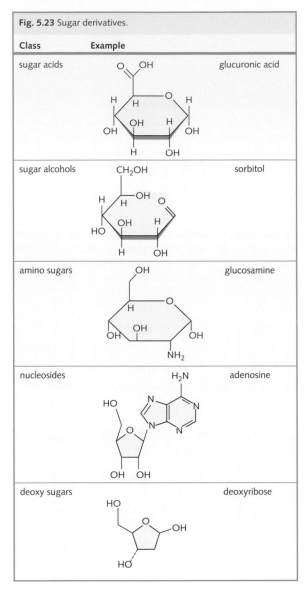

Fig. 5.23 Sugar derivatives.

- nucleosides: specialized amino sugars, in which the amino residue is a pyrimidine or a purine – e.g. adenosine, deoxyadenosine, thymidine and deoxythymidine
- deoxy sugars: one hydroxyl group is substituted for hydrogen – e.g. deoxyribose and fucose.

Basic molecular biology and genetics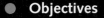

Objectives

By the end of this chapter you should be able to:
- Discuss broadly the biosynthesis of purines and pyrimidines, and identify the rate-limiting steps of each.
- Compare and contrast the structures and functions of the different RNA molecules.
- Draw a diagram to show the stages of the cell cycle and explain the role of cyclins in the regulation of the cell cycle.
- Describe the different orders of DNA packaging, from DNA to chromosome.
- Draw an annotated diagram of a eukaryotic replication fork, and contrast it with that of a prokaryotic replication fork.
- Understand the process of eukaryotic transcription and be able to compare and contrast with prokaryotic transcription.
- Summarize the salient features of initiation, elongation and termination of eukaryotic protein synthesis.
- Define mitosis and meiosis, list the stages of each and highlight the differences between them.
- Identify repair mechanisms required to repair single-stranded DNA damage and double-stranded DNA damage.

ORGANIZATION OF THE CELL NUCLEUS

The nucleus is the largest eukaryotic cell structure. The nucleoplasm is in constant contact with the cell cytoplasm via pores in the nuclear membrane. The nucleus consists of DNA (deoxyribonucleic acid), proteins and RNA (ribonucleic acid), and plays a vital role in:

- protein synthesis (see p. 60)
- the passage of genetic information from one generation to the next (see p. 78).

Structures of the nucleus

Nuclear envelope

The nuclear envelope encloses the nucleus. It consists of two layers of membrane, the outer continuous with the endoplasmic reticulum (Fig. 6.1). The periplasm, the space between the inner and outer membranes, forms a continuum with the lumen of the endoplasmic reticulum (ER). Ribosomes are attached to both the outer layer of the nuclear envelope and the ER.

Nuclear pores

Nuclear pores are found at points of contact between the inner and outer membranes (Fig. 6.2). They are electron-dense structures consisting of eight protein complexes arranged around a central granule. They control the passage of metabolites, macromolecules and RNA subunits between the nucleus and the cytoplasm. Molecules up to 60 kDa pass freely through these pores, but transport of larger molecules is ATP-dependent and requires the receptor-mediated recognition of a nuclear targeting sequence by the pore complex.

Nucleoli

Nucleoli are dense structures in the nucleus where ribosomal RNA synthesis and ribosome assembly occurs. There may be one, several, or no visible nucleoli in a nucleus (see p. 15).

HINTS AND TIPS

Cells inactive in protein synthesis tend to have heterochromatin rich nuclei and no nucleoli.

Nuclear matrix

The nuclear matrix consists of DNA, nucleoproteins and structural proteins. Nucleoproteins are closely associated with DNA and are defined as histone or non-histone. Histones are strongly basic, globular proteins around which DNA winds in a regular fashion, to form chromatin (see p. 15). Chromatins interact with the lamins, a type of intermediate filament, which are

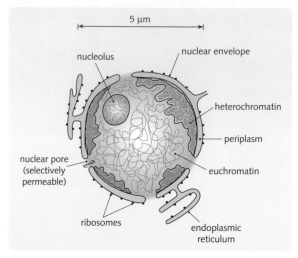

Fig. 6.1 Structure of the nucleus.

arranged in a lattice forming a thin shell underlying the inner nuclear membrane. A less regularly organized network of intermediate filaments surrounds the outer membrane, that together provide mechanical support for the nuclear envelope.

NUCLEIC ACIDS

Nucleic acids are produced from nucleotide polymerization (Fig. 6.3). During synthesis a series of nucleic acid condensation reactions occur between phosphate and sugar groups, producing strong phosphodiester bonds. Long, unbranching chains form with linkages between C3 and C5 of each sugar, so $5'$–$3'$ or $3'$–$5'$ is used to describe the orientation of nucleotides in a chain. Each sugar is separated from the next by a phosphate group, forming a strong and rigid sugar–phosphate

Fig. 6.2 Nuclear pore structure. (Adapted from Stevens and Lowe, 1997.)

backbone from which the bases project (Fig. 6.4). DNA and RNA are nucleic acids with an integral role in all living cells.

Nucleotides and nucleosides

A nucleotide is a compound of a pentose sugar residue (deoxyribose in DNA, ribose in RNA; Fig. 6.5) attached to a base (purine or pyrimidine) and a phosphate group (see Fig. 6.3). They are the subunits of nucleic acids and are named according to whether the base is a ribonucleotide or a deoxyribonucleotide and the number of attached phosphate groups (e.g. adenosine monophosphate).

HINTS AND TIPS

A nucleoTide has Three moieties.

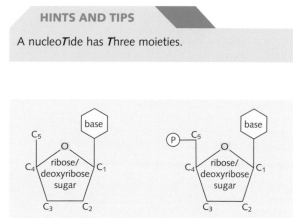

base+sugar=nucleoside   base+sugar+phosphate=nucleotide

Fig. 6.3 Nucleoside and nucleotide structure. Deoxyribose has an H group on C_2 of the ribose moiety, whereas ribose has an OH group at this position. (Adapted from Molecular Biology of the Cell, 3rd edn, by B Alberts et al, Garland Publishing, 1994. Reproduced by permission of Routledge, Inc., part of the Taylor & Francis Group.)

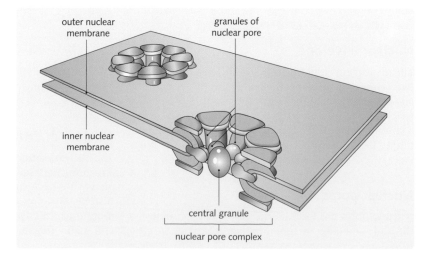

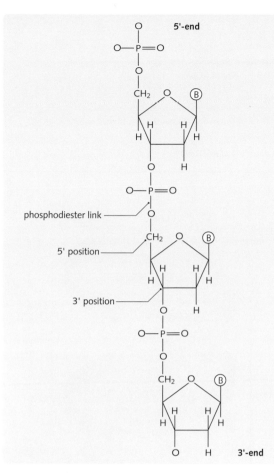

Fig. 6.4 Nucleic acids. B indicates the position of the base. (Adapted from Norman and Lodwick, 1999.)

Fig. 6.5 Comparing DNA and RNA.

Feature	DNA	RNA
Sugar	Deoxyribose	Ribose
Base pairing	A–T/G–C	A–U/G–C
Structure	Double helix	Single stranded structures

A nucleoside is a compound of a sugar residue and a base linked by an N-glycosidic bond between C1 of the sugar and an N atom of the base (see Fig. 6.3). The naming of nucleosides is similar to that of nucleotides.

Purines and pyrimidines

The nucleotide bases in nucleic acids are heterocyclic molecules derived from either purines (adenine, guanine) or pyrimidines (cytosine, thymine, uracil). Uracil is not present in DNA, but replaces thymine in RNA. They are planar aromatic rings that contain nitrogen.

Almost all eukaryotic cells are capable of purines and pyrimidine synthesis *de novo*, reflecting an essential role in cell survival.

HINTS AND TIPS

Pyrimidines are **CUT** from purines. The pyrimidines are: Cytosine (DNA and RNA), Uracil (RNA), Thiamine (DNA), and 'single-ringed', hence **CUT** from the 'double-ringed' purines.

Purine biosynthesis

Purines are synthesized in an 11-stage pathway, starting with the formation of 5-phosphoribosyl-1-pyrophosphate (PRPP). Through this, ATP, folate (tetrahydrofolate) derivatives, glutamine, glycine and aspartate are utilized to yield inosine monophosphate (IMP), which is converted to adenosine monophosphate (AMP) or guanosine monophosphate (GMP). A complex negative feedback control network operates to prevent excessive AMP and GMP build-up, the rate-limiting step lying in the first reaction (the synthesis of PRPP by PRPP synthetase).

Clinical Note

Gout results from abnormal catabolism of purines, either from an excess of purines or partial deficiency in hypoxanthine guanine phosphoribosyl transferase (HGPRT). Clinical manifestations of abnormal purine catabolism arise from the insolubility of the degradation by-product, uric acid, most of which is usually excreted via the kidneys. Increased formation or reduced excretion of uric acid leads to hyperuricaemia and the formation of sodium urate crystals, which precipitate in the synovial fluid of the joints leading to severe inflammation and arthritis.

Pyrimidine biosynthesis

Pyrimidine synthesis involves a six-step pathway, in which the ribose sugar is incorporated as one of the final steps. From a starting point of carbamoyl phosphate, derived from glutamine and bicarbonate, uridine monophosphate (UMP) is formed. UMP is then phosphorylated to uridine triphosphate (UTP), where it is free to be aminated to form cytidine triphosphate (CTP). Uridine nucleotides are also the precursors for *de novo* synthesis of the thymine nucleotides.

Fig. 6.6 Major salvage pathways for pyrimidine and purine bases.

	Purines	Pyrimidines
Major bases salvaged	Hypoxanthine, guanine	Uracil (thymine)
Enzymes involved	Hypoxanthine-guanine-phospho-ribosyltransferase	Uridine phosphorylase, uridine kinase (thymidine phosphorylase, thymidine kinase)
Products	IMP, GMP	UMP (dTMP)

Salvage pathways

In addition to *de novo* synthesis, nucleotides can also be synthesized from the breakdown of endogenous nucleic acids through salvage pathways (Fig. 6.6), in which preformed bases are recovered and reconnected to a ribose unit.

Pentose sugars

These are five-carbon rings (see Fig. 6.4). Deoxyribonucleotides are formed by the reduction of the ribose group of the corresponding ribonucleotide (see p. 74).

DNA double helix

Discovered by Watson and Crick in 1953, the structure of DNA is one of two intertwined polynucleotide strands held together by base pairing to form a double helix. Adenine and thymine pair via two hydrogen bonds between opposing strands, whereas guanine and cytosine pair via three hydrogen bonds. Base pairing results in two complementary polynucleotides, which run antiparallel to each other (i.e. one runs 5′–3′, the other runs 3′–5′; Fig. 6.7).

RNA molecules

In eukaryotes, all single-stranded RNA is produced from DNA by transcription, predominantly in the nucleus. It moves out into the cytoplasm to function.

Messenger RNA

Messenger RNA (mRNA) carries genetic information from the nucleus into the cytoplasm. In eukaryotes, it is derived by splicing the initial RNA transcript (heteronuclear RNA). This single-stranded mRNA forms the template upon which polypeptides are manufactured during translation (Fig. 6.8).

Heterogeneous nuclear RNA

Heterogeneous nuclear RNA (hnRNA) is the primary mRNA transcript produced by eukaryotic cells. It is very short lived, as it is processed into mature mRNA. Unlike

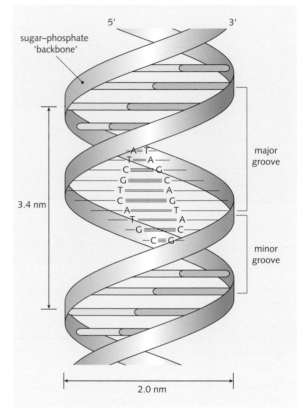

Fig. 6.7 The DNA double helix is a right-handed helix with a common axis for both strands. There are ten base pairs per turn.

the final mRNA transcript it contains introns, which are subsequently removed by RNA splicing (see p. 88).

Transfer RNA

Transfer RNA (tRNA):

- carries specific amino acids to the site of protein synthesis
- has two active sites allowing it to carry out its functions
- is a linear molecule with an average of 76 nucleotides
- exhibits extensive intramolecular base pairing, giving it a 'clover-leaf'-shaped secondary structure.

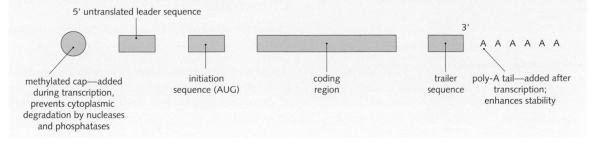

5' untranslated leader sequence

3'

A A A A A A

methylated cap—added during transcription, prevents cytoplasmic degradation by nucleases and phosphatases

initiation sequence (AUG)

coding region

trailer sequence

poly-A tail—added after transcription; enhances stability

Fig. 6.8 Structure of eukaryotic mRNA.

Some tRNA bases undergo post-translational modifications, thought to be required for tRNA–protein interactions or tRNA molecule stabilization. Figure 6.9 shows tRNA secondary structure.

Ribosomal RNA

Ribosomal RNA (rRNA) is a ribosomal component. In a eukaryotic cell each ribosome consists of two unequal subunits, made up of proteins and RNA, called the S (small) and L (large) subunits (Fig. 6.10). The RNA molecules undergo extensive intramolecular base pairing, which determines the ribosomal structure. Ribosomes can self-assemble under physiological conditions with the correct complement of components.

DNA PACKAGING AND CHROMOSOMES

DNA is found predominantly in the nucleus, but also in mitochondria (see p. 90). It is a template during transcription, and the vehicle of inheritance (i.e. passed from one generation to the next). In the nucleus of a normal human cell, there are 46 chromosomes each containing 48–240 million bases of DNA. Watson and Crick's double helix model predicts that each chromosome would have a contour length of 1.6–8.2 cm and that the total length of the DNA would be about 3 m. However, the average nucleus has a diameter of approximately 5 µm so a high degree of organization is needed to fit this amount of DNA into the nucleus (Fig. 6.11).

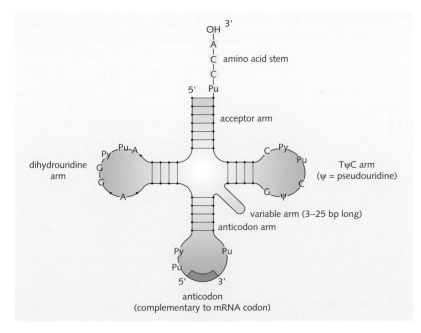

3'
OH
A
C   amino acid stem
C
5'   Pu

acceptor arm

dihydrouridine arm

Pu–A
Py
G
G
A

C—Py
Pu
C
C
G—ψ

TψC arm
(ψ = pseudouridine)

variable arm (3–25 bp long)

anticodon arm

Py   Pu
Pu

5'   3'

anticodon
(complementary to mRNA codon)

Fig. 6.9 Secondary structure of tRNA, consisting of five arms. The active sites are on the acceptor arm, where the 3' terminal CCA group can accept a specific amino acid, and the anticodon is on the anticodon arm, which recognizes the corresponding mRNA codon. Specific base pairing within the five arms helps to maintain the secondary structure. A, adenine; C, cytosine; G, guanine; Pu, purine; Py, pyrimidine; U, uridine.

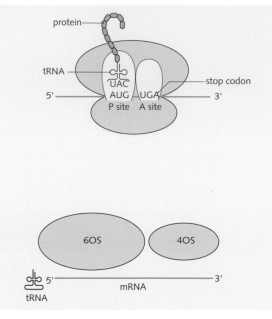

Fig. 6.10 Structure of a eukaryotic ribosome. Ribosomes consist of two unequal subunits, composed of RNA and protein, held together by magnesium ions. The ribosome has binding sites for mRNA, the peptidyl tRNA (P-site) and aminoacyl tRNA (A-site). S, Svedberg unit of sedimentation. (Adapted from Baynes and Dominiczak, 1999.)

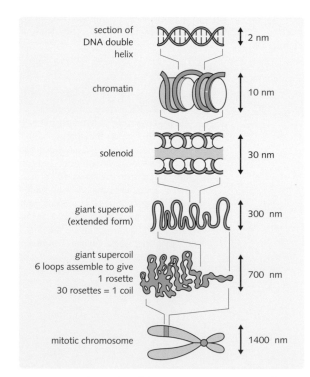

Fig. 6.11 Higher order of chromatin structure. The different levels of DNA condensation, from DNA double helix to mitotic chromosome.

Chromatin

Chromatin is the collective name for the long strands of DNA, RNA and their associated nucleoproteins. During interphase of the cell cycle (see p. 94), chromatin is dispersed throughout the nucleus, becoming more compact during mitosis or meiosis (see p. 97). Two types of chromatin can be seen with electron microscopy (see Ch. 2).

1. Heterochromatin – which is electron dense and distributed around the periphery of the nucleus and in discrete masses within the nucleus. The DNA is in close association with nucleoproteins, and it is not active in RNA synthesis.
2. Euchromatin – which is electron lucent and represents DNA that is actually or potentially active in RNA synthesis.

In the progressive levels of chromosome packaging:

- DNA winds onto nucleosome spools
- the nucleosome chain coils into a solenoid
- the solenoid forms loops, which attach to a central scaffold
- the scaffold plus loops arrange into a giant supercoil.

Nucleosomes

A nucleosome is formed by 146 bp of DNA wound twice around an octamer of histone proteins (Fig. 6.12). The octamer consists of two copies each of the histone proteins H2A, H2B, H3, and H4. Histone proteins are conserved throughout eukaryotic evolution. They contain a high proportion of positively charged amino-acid residues that can form ionic bonds with the negatively charged DNA. This interaction does not depend on DNA sequence and theoretically histones can bind with any piece of DNA. However, *in vivo*, the position of histone binding is influenced by:

- AT content (bends more easily than GC)
- the presence of other tightly bound proteins.

Nucleosome bound regions of DNA are separated by a region of linker DNA that varies from 0–80 bp in length. Consequently, on electron micrographs nucleosomes appear as 11-nm 'beads' on a 2-nm DNA 'string'.

DNaseI is an endonuclease that breaks the internal phosphodiester bonds in DNA, irrespective of its base sequence. The regulatory regions of genes are frequently bound by proteins that prevent histones from binding. Since histone binding protects DNA from degradation with DNaseI, the regulatory regions of genes are particularly sensitive to this enzyme, and they are sometimes called 'nuclease-hypersensitive sites'.

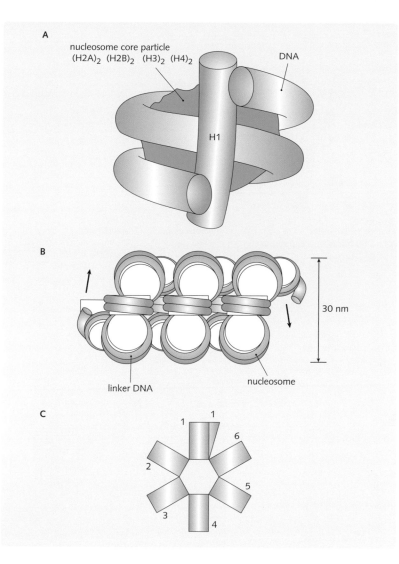

Fig. 6.12 Chromatin fibre organization. (A) The nucleosome core particle is composed of pairs of histones. 166 base pairs of DNA wind around each nucleosome. Linker DNA consisting of 8–114 base pairs runs between one nucleosome and the next. (B) Chromatin consists of nucleosomes bound together through their H1 proteins (not shown in this part of the figure). (C) Bound nucleosomes form a solenoid, with six nucleosomes per turn. (Adapted with permission from Molecular Biology of the Cell, 3rd edn, by B Alberts et al, Garland Publishing, 1994. Reproduced with permission of Routledge, Inc., part of the Taylor & Francis Group.)

Solenoid formation

The second level of DNA packing is mediated by histone HI, binding together adjacent nucleosomes to condense DNA into the supercoiled 30-nm fibre, which is also called the solenoid (see Figs 6.11, 6.12). The solenoid exhibits six to eight nucleosomes per turn of the spiral, corresponding to heterochromatin.

Giant supercoil

The third level of organization is thought to involve the formation of transcriptional units of DNA loops radiating from a central scaffold of non-histone proteins.

Chromosomes

At metaphase (see p. 97), chromatin is maximally condensed and forms 1400-nm fibres. After cell staining these structures are visible as chromosomes under light microscopy.

Centromeres

Each metaphase chromosome is composed of two identical sister chromatids. Chromatids are connected at a central region called the centromere, above and below which chromatin strands loop across and between chromatids to hold them together (Fig. 6.13). Centromeres consist of hundreds of kilobases of repetitive DNA and

Fig. 6.13 Anatomy of a chromosome showing the three shapes: metacentric, submetacentric and acrocentric.

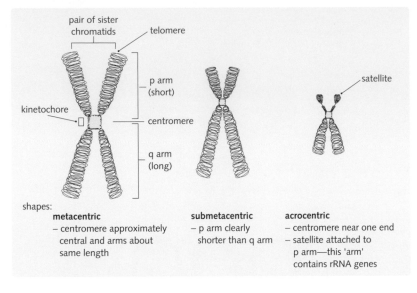

pair of sister chromatids

telomere

kinetochore

p arm (short)

satellite

centromere

q arm (long)

shapes:

metacentric
– centromere approximately central and arms about same length

submetacentric
– p arm clearly shorter than q arm

acrocentric
– centromere near one end
– satellite attached to p arm—this 'arm' contains rRNA genes

are responsible for the movement of chromosomes at cell division. Each centromere divides the chromosome into short (p) and long (q) arms.

> **HINTS AND TIPS**
>
> Arms of the chromosome, remember: p is petite (short arm), queues are long (long arm).

Centromere position can be used to categorize chromosomes morphologically (see Fig. 6.13).

- Acrocentric – centromeres located very close to one end, yielding a small short arm, often associated with small pieces of DNA called satellites, encoding rRNA.
- Metacentric – centromeres located in the middle, yielding arms of roughly equal length.
- Submetacentric – off-centre centromere so that one arm is longer than the other.

The kinetochore is an organelle located at the centromere region (Fig. 6.14). It acts as a microtubule organizing centre and facilitates spindle formation by polymerization of tubulin dimers to form microtubules early in mitosis (see Ch. 4).

Telomeres

The ends of chromosomes are protected by DNA structures called telomeres. Telomeres are tandem repeats of the hexameric sequence 'TTAGGG', ending in a 3' single-stranded overhang that ranges in length from about 50–400 nucleotides and loops back on itself to form

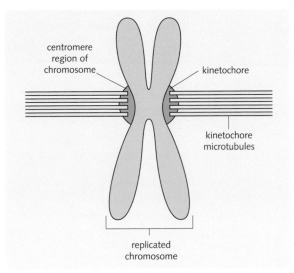

centromere region of chromosome

kinetochore

kinetochore microtubules

replicated chromosome

Fig. 6.14 The kinetochore. The kinetochore contains two regions: an inner kinetochore, which is tightly associated with the centromere DNA, and an outer kinetochore, which interacts with microtubules.

the T-loop (Fig. 6.15). Telomeres have several functions in preserving chromosome stability, including:

- preventing abnormal end-to-end fusion of chromosomes
- protecting the ends of chromosomes from degradation
- ensuring complete DNA replication
- having a role in chromosome pairing during meiosis.

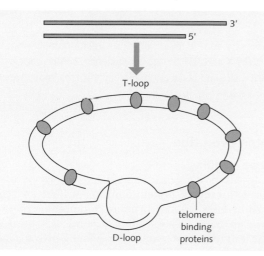

Fig. 6.15 Telomeres. T-loop formation prevents telomerase activity, blocking excessive telomere extension. With each cell division, the telomeric DNA shortens, until the telomere is too short to support a T-loop structure. At this point telomerase can act, lengthening the telomere, allowing the T-loop to reform. This cycle of shortening and lengthening means the T-loop effectively protects the telomere end.

DNA polymerases require an RNA primer (see p. 86). This poses a potential problem at the ends of eukaryotic chromosomes because there is nowhere for the lagging strand primer to bind to facilitate replication of the terminal sequence. Thus, potentially the chromosome could become progressively shorter after successive rounds of replication, resulting in a loss of genetic information. In order to avoid this, the cell employs the addition of non-coding DNA to the end of the 3' tail via the enzyme telomerase (see p. 86).

Nuclear genes

Genes are sequences of DNA that encode proteins, and are composed of a transcribed region and a regulatory sequence. The 'one gene, one polypeptide' hypothesis states that the base sequence of DNA determines the amino-acid sequence in a single corresponding polypeptide. By convention, gene sequences are described in the direction 5'–3', as this is the direction of *in vivo* nucleic acid synthesis.

Genes lie within expanses of 'non-coding DNA' which, until recently, were not believed to serve any function. However, research suggests that 50% of non-coding DNA is transcribed and produces non-coding RNA (ncRNA) possibly with a critical role in DNA structure regulation, RNA expression and protein translation and functions.

Eukaryotic gene structure

The typical eukaryotic gene consists of a number of conserved features (Fig. 6.16).

Promoters

The promoter region lies in the 5' DNA immediately preceding a gene and is sometimes called the 'upstream flanking region'. 'Consensus' sequences have been found in both prokaryotes and eukaryotes and are vital for promoter function. The most conserved sequence in the *E. coli* promoter is the TATAAT box, which forms the initial binding site for the transcription enzyme RNAP (see Ch. 1). The situation is more complex in eukaryotes, where multiple consensus sequences have been detected, including the GC box (5'-GGGCGGG-3'); the TATA box (5'-TATAAAAA-3') and the CAAT box (5'-GGCCAATCT-3'), which act as binding sites for specific transcription factors.

Eukaryotic genes may also require enhancer sequences that may be situated many kilobases from the start of transcription (see p. 87).

Introns and exons

The typical eukaryotic gene contains both exons and introns (Fig. 6.16). Exons are transcribed sequences (i.e. represented in final mRNA) – the majority of exons will code for amino acids, but 5' or 3' UTRs are also included in exons. Introns are the non-coding intervening sequences. The length of a typical exon is a few hundred

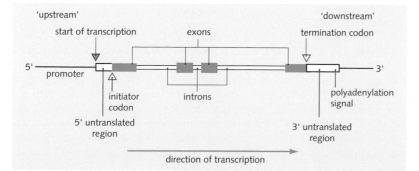

Fig. 6.16 Structure of a typical eukaryotic gene. The first and final exons include sequence that is transcribed and present in the mature mRNA, but not translated. These are called the 5'- and 3'-UTR (untranslated regions) respectively. This diagram is not to scale and in reality the average intron is much longer than the average exon. (Adapted from Nussbaum, McInnes and Willard, 2001.)

base pairs whereas introns tend to be several kilobases long. Introns are:

- rare in prokaryotes
- uncommon in lower eukaryotes, such as yeast
- abundant in higher eukaryotes (e.g. vertebrate structural genes).

3' sequences

The 3'UTR is defined as the sequence extending from a coding region stop codon up to the point at its 3' end where the transcript is cleaved, and is thought to influence mRNA translation, localization and stability. In particular, the polyadenylation signal (5'-AATAAA-3') determines the site of polyadenylation of the resultant mRNA molecule, which in turn protects the molecule from the action of enzymes (exonucleases) and is important for transcription termination, export of mRNA from the nucleus and translation.

Active chromatin

If nuclei isolated from different vertebrate cell types are treated with the enzyme DNaseI, they show different degradation patterns. This is due to different actively transcribed genes being more sensitive to the enzyme, suggesting that active chromatin and inactive chromatin are packaged differently. In contrast to inactive chromatin, active chromatin has:

- less tightly bound histone H1
- highly acetylated nucleosomal histones
- less phosphorylated histone H2B
- enrichment of a H2A variant.

The differences between active and inactive chromatin suggest that chromatin structure is important in the regulation of gene expression. Large areas of the genome may be transcriptionally silenced during differentiation due to specialized packing. Methylation status may influence chromatin structure.

Methylation

The CpG dinucleotide (i.e. C followed by G, not to be confused with C–G base pair) is generally underrepresented in the genome. However, it is found at expected levels in promoter regions. In the promoters of inactive genes the 5' position of cytosine is frequently methylated, whereas it is generally unmethylated in active genes. Methyl-CpG binding proteins bind to methylated regions and recruit histone deacetylase, which may trigger inactive chromatin formation. In this way, methylation may act to mark chromatin for inactivation, although the factors that prompt it to do so are unclear. However, methylation occurs due to the replication of methylation patterns by maintenance methylase, a methyl group added to a previously hemimethylated site.

Gene evolution

Intron shuffling

Only eukaryotic genes contain introns. Exons often encode functional protein domains. If these are separated by long introns, then any chromosome break is likely to be in a non-coding region. Such breakage may lead to recombinational exchange between genes, 'shuffling' new combinations of exons together, to produce new, potentially advantageous proteins. If this process were to occur in the absence of introns, it is likely that the new protein would be out of frame, losing functional domains.

Gene duplication and multigene families

Gene duplication is a rare consequence of normal recombinational events. Duplicated genes are free to mutate and evolve new functions and patterns of expression, since they are not required for survival. The mechanism of gene duplication events means that some functionally related proteins are clustered close together on the same chromosome.

HLA complex

The human leukocyte antigen (HLA) complex (a.k.a. the major histocompatibility complex, MHC) on the short arm of chromosome 6 is an example of a gene cluster. The HLA genes:

- are mostly members of the immunoglobulin superfamily – hundreds of genes involved in cell surface recognition
- are characterized by the presence of immunoglobulin 'domains' consisting of 110 amino-acid residues stabilized by a disulphide bridge
- share sets of related exons evolved from an ancestral gene by duplication.

There are three classes of HLA genes (Fig. 6.17A). Classes I and II encode proteins that are involved in the presentation of peptide antigens to T cells. These are usually foreign antigens derived from infective pathogens but also include 'non-self' antigens introduced through organ transplantation (Fig. 6.17B). Class III genes encode proteins that modulate or regulate immune responses in other ways, including tumour necrosis factor (TNF) and complement proteins (C2, C4) (Fig. 6.17B).

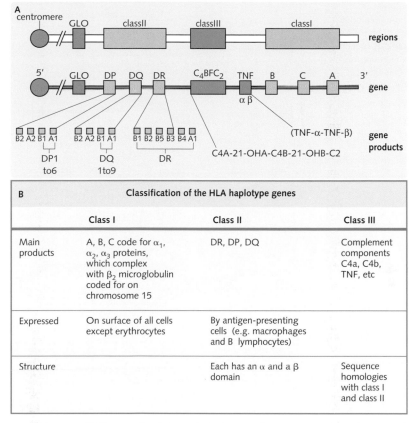

Fig. 6.17   (A) The major histocompatibility complex (HLA region), showing the regions, genes and gene products on the short arm of chromosome 6 (6p21.1 to p21.3). GLO, glyoxalase; TNF, tumour necrosis factor. (B) Features of different classes of HLA genes. (Adapted from Kumar and Clark, 2005.)

The term haplotype describes a cluster of alleles that occur together on a DNA segment and are inherited together. Children get one HLA haplotype from their mother and one HLA haplotype from their father. In the case of renal transplantation, the degree of matching for HLA loci A, B, and DR (Fig. 6.17A) is assessed to reduce the risk of organ rejection.

Pseudogenes

Pseudogenes are DNA sequences resembling structural genes but cannot be translated into a functional protein. Their presence within the human genome gives an insight to gene evolution, especially as many functional genes have pseudogene equivalents. They are thought to have arisen in a number of ways, including incomplete duplication events, silencing of functionally redundant genes and the insertion of complementary DNA sequences produced by the action of reverse transcriptase on a naturally occurring mRNA transcript (retrotransposition).

DNA REPLICATION

As discussed in Chapter 1, DNA replication results in duplication of nuclear DNA in preparation for cell division. Replication proceeds in the 5'–3' direction with new nucleotides being attached to the 3' OH of the growing molecule. One strand, known as the leading strand, is formed continuously, moving in the direction of the replication fork. The other strand, the lagging strand, is formed in short sequences of 1000–5000 nucleotides known as Okazaki fragments, which are then joined enzymatically (see Fig. 6.20).

Eukaryotic DNA replication

Eukaryotic DNA synthesis is remarkably similar to that seen in prokaryotes (see Ch. 1). However, eukaryotes have many more origins of replication (Fig. 6.18), which are activated simultaneously during the S phase of the cell cycle (see p. 93), enabling rapid replication

Fig. 6.18 Eukaryotic DNA replication. There are multiple origins of replication, but replication is initiated at specific points at specific times to ensure that the entire genome is replicated only once. (Adapted from Jorde et al, 1997.)

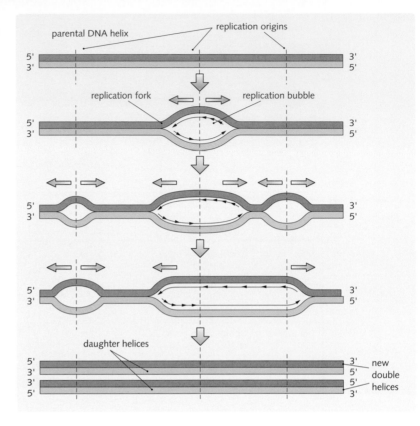

of entire chromosomes. The origin recognition complex, along with licensing factors, triggers an origin of replication to begin DNA replication, and prevents over-replication. Replication proceeds in two directions from each origin of replication and continues until neighbouring forks fuse. The rate of DNA synthesis in eukaryotes is slower than in prokaryotes cells.

Heterochromatin replicates later in the S phase than euchromatin. The genome is replicated once only, and chromatin may be marked after replication to prevent it being replicated again, possibly by DNA methylation.

DNA polymerases

Over 19 polymerases have been identified in eukaryotes and α, β, γ, δ and ε are the key enzymes required to maintain genome integrity. Of these, two (α and δ) are especially important for chromosome replication.

Replication forks

After the replication fork passes, adding new histones reforms chromatin structure. Histone proteins in eukaryotic chromosomes may account for the slower

eukaryotic replication and shorter Okazaki fragments than prokaryotic replication (Fig. 6.19).

Additional proteins involved in DNA replication

Proliferating cell nuclear antigen

Proliferating cell nuclear antigen (PCNA) is the eukaryotic counterpart of the regulated sliding clamp protein of *E. coli*. (Fig. 6.20). It acts as a co-factor for DNA polymerase δ in S phase and also during DNA synthesis associated with DNA damage repair mechanisms.

Replication protein A

Replication protein A (RPA) is the eukaryotic equivalent to the single-strand binding proteins. RPA molecules facilitate the unwinding of the helix to create two replication forks. Experiments *in vitro* have shown that replication is 100 times faster when these proteins are attached to the single-stranded DNA. RPA is also required for nucleotide excision repair, and homologous recombination (Fig. 6.20).

Fig. 6.19 Prokaryotic and eukaryotic DNA replication.

	E.coli (prokaryote)	Mammalian (eukaryote)
Site	Cytoplasm	Nucleus (and mitochondrion)
No. of proteins involved	30	100s
DNA polymerase	Pol I—fidelity and repair Pol III—DNA synthesis	Four enzymes identified, α polymerase is the principal nuclear polymerase
Initiation	Single origin of replication (OriC)	Multiple origins of replication (spatially and temporally separated during DNA replication)
Rate of replication	10^3 nucleotides/s	10^2 nucleotides/s
Post-replication	RNA primers removed from lag strand by Pol I 5'–3' exonuclease	RNA primers removed by 5'–3' exonuclease (NOT associated with α polymerase)
Timing of replication	Continuous DNA synthesis between cell divisions	DNA synthesis and cell division separated by G_1 and G_2 (gap) phases
Okazaki fragments	Large (1000 – 2000 bp)	Small (100 – 200 bp)
DNA polymerase	Same for leading and lagging strands	Different for leading and lagging strands

---

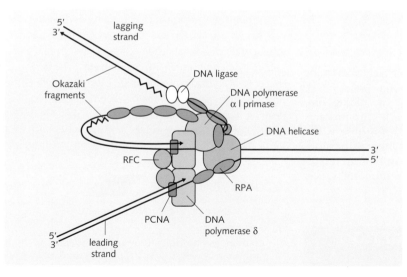

Fig. 6.20 Eukaryotic DNA replication. Leading strand synthesis starts with the primase activity of DNA polymerase α to lay down an RNA primer, to which it adds a stretch of DNA. RFC assembles PCNA at the end of the primer, followed by displacement of DNA polymerase α. DNA polymerase δ binds to PCNA at the 3' ends of the growing DNA strand to carry out highly processive DNA synthesis. Lagging strand synthesis begins in the same manner as leading strand synthesis. RNA primers are synthesized by DNA polymerase α every 50 nucleotides. Polymerase switching generates Okazaki fragments, which are ultimately ligated together by DNA ligase.

Replication factor C

Replication factor C (RFC) is required, in the presence of ATP, to load the PCNA sliding clamp onto DNA, thereby recruiting DNA polymerases to the site of DNA synthesis (Fig. 6.20).

Topisomerases

These are enzymes responsible for unwinding and winding DNA. Topisomerases have a role in both DNA replication and protein synthesis. This enzyme ensures the double helix is unwound in order to correctly access the information stored as the two DNA strands are intertwined.

Leading strand synthesis

Progresses (Fig. 6.20) as follows:

- DNA polymerase α has primase activity which lays down an RNA primer
- DNA polymerase α adds a stretch of DNA to the RNA primer
- RFC assembles PCNA at the end of the primer
- PCNA displaces DNA polymerase α
- DNA polymerase δ binds to PCNA to carry out highly processive DNA synthesis.

Lagging strand synthesis

Lagging strand synthesis (Fig. 6.20) commences in a similar fashion to leading strand synthesis:

- DNA polymerase α synthesizes RNA primers every 50–75 nucleotides
- PCNA mediated polymerase switching extends the RNA–DNA primers to generate Okazaki fragments
- DNA polymerase δ extends its activity towards the RNA primer of the downstream Okazaki fragment, at which point enzymes (RNases and exonucleases) remove the RNA primer
- DNA polymerase δ fills in the gap as the RNA primer is being removed
- DNA ligase joins the Okazaki fragment to the growing strand.

Telomerase

Telomeres are predisposed to progressively shortening with each round of DNA replication (see p. 81). Telomerase circumvents shortening by adding protective sequences to the ends of each chromosome. Although active in germ-line cells, telomerase is not normally active in somatic cells. The progressive shortening of telomeres in somatic tissue may be an important component of ageing. Moreover, immortal cancer cells frequently show regained telomerase activity.

EUKARYOTIC TRANSCRIPTION AND RNA SYNTHESIS

Definition of transcription

Transcription is RNA synthesis according to a DNA template. It is catalysed by DNA-dependent RNA polymerases, which unwind dsDNA, exposing unpaired bases upon which DNA–RNA hybrids form. RNA is synthesized 5′–3′. The DNA template displays polarity as only one strand can act as a template (template strand). The non-transcribed strand, the coding strand, has the same base composition as the RNA except that thymines (Ts) are substituted for uracils (Us).

The template strand is transcribed. It is identified by RNA polymerase, which binds to the specific DNA sequences that comprise the promoter. Transcription may be influenced and regulated by both *cis* and *trans* acting factors:

- *cis* acting factors are specific sequences of DNA that lie on the same molecule of DNA as the gene they regulate
- *trans* acting factors are proteins that bind to *cis* acting elements. They are transcribed from genes distinct from the ones they regulate.

Eukaryotic transcription

Eukaryotes have three chromosomally encoded RNA polymerases, which recognize different promoters and transcribe different types of RNA molecules. The TATA (Hogness–Goldberg) box is found in the promoters of genes transcribed by RNA polymerase II. These polymerases can be identified because they differ in their sensitivity to α-amanitin toxin (Fig. 6.21).

Clinical Note

Gilbert syndrome (GS) is a benign, mildly symptomatic, unconjugated hyperbilirubinaemia, and affects ~10% of the population. Bilirubin UDP glucuronosyltransferase 1 (UGT1A1) activity in patients with GS is decreased to about 30% of normal, leading to a failure of uptake of albumin-bound bilirubin into hepatocytes. The large majority of mutations found to be associated with GS involves an expansion of the TATA box in UGT1A1 leading to reduced expression and, ultimately, reduced activity of the enzyme product.

HINTS AND TIPS

There are three different RNA polymerases in eukaryotes:
1. I transcribes ribosomal RNA in the nucleolus
2. II transcribes mRNA in the nucleoplasm
3. III transcribes tRNA (and one rRNA species) in the nucleoplasm.

Sequence of events

Initiation

Eukaryotic transcription is more complicated than in prokaryotes, requiring the presence of several transcription factors (proteins required to initiate or regulate eukaryotic transcription). All genes that are transcribed and

Fig. 6.21 Eukaryotic RNA polymerases. These differ with respect to their template specificity, localization and sensitivity to α-amanitin. (Adapted from Biochemistry, 3rd edn, by L Stryer, W.H. Freeman and Company, 1988.)

RNA polymerase	Localization	Cellular transcripts	Effect of α-amanitin
I	Nucleolus	18S, 5.8S, and 28S rRNA	Insensitive
II	Nucleoplasm	mRNA precursors and hnRNA	Strongly inhibited
III	Nucleoplasm	tRNA and 5S rRNA	Inhibited by high concentrations

expressed via mRNA are transcribed by the RNA polymerase II complex. It contains at least six basal transcription factors (TFII) A, B, D, E, F and H, and may be assembled, and transcription initiated, as follows (Fig. 6.22).

- TFIID (TBP (TATA binding protein) and numerous TAFs (TBP associated factors)) recognizes and binds to the TATA box.
- TFIIA binds TFIID and DNA, stabilizing the interaction.
- TFIIB binds to TFIID, recruiting TFIIF-RNA Pol II.
- TFIIF, having helicase activity, may help expose the DNA template strand.
- TFIIE binds and recruits TFIIH to the complex.
- TFIIH phosphorylates Pol II.
- Pol II is released from the complex and begins transcription.

If the transcription complex contains only the six basal transcription factors highlighted above, the level of transcription is low. Physiological levels of transcription require enhancers. These are *cis* acting elements that may be many kilobases away from the start of transcription in either direction. Enhancers are bound by *trans* acting transactivating proteins, which are then able to associate with the polymerase complex to increase the level of transcription (Fig. 6.23).

Since the expression of transactivators is tissue specific, enhancers facilitate tissue-specific control of gene expression in multicellular organisms. Similarly, transcription from a gene may be reduced in a tissue-specific manner by the presence of silencers (the *cis* acting element) and repressor proteins (the *trans* acting element).

Elongation

The overall equation is the same as for prokaryotic elongation:

$$(RNA)_n + XTP \rightarrow (RNA)_{n+1} + PP_i \rightarrow 2P_i$$

However, unlike prokaryotes, the transcript is modified while it is still being transcribed by the addition of a 5′ cap (see below).

The length of the primary RNA transcript (hnRNA), at an average of 7000 nucleotides, is very much larger than the average prokaryotic transcript. Moreover, it is larger than the predicted 1200 nucleotides needed to code for an average protein of 400 amino-acid residues. This reflects the presence of introns, which are not found in prokaryotic transcripts.

Termination

Unlike prokaryotes, eukaryotic genes have no strong termination sequences. Instead, RNA polymerase II continues transcribing up to 1000 to 2000 nucleotides beyond where the 3′ end of the mature mRNA will be. The actual 3′ end is determined by the cleavage of the transcript at a highly conserved AAUAAA sequence, potentially generating the 3′ end of the message to which a poly(A) tail is added (see p. 89). This sequence is thought to be important as, when the corresponding DNA template sequence occurs, no mature mRNA is made.

Eukaryotic post-transcriptional modification

Addition of a 5′ cap

This is a very early modification that occurs soon after transcription initiation. The cap structure, a 7-methylguanosine residue, is enzymatically added to the 5′ end of the hnRNA molecule by a unique 5′–5′ linkage (Fig. 6.24). It is thought to have four main functions:

1. It protects the mRNA from enzymatic attack
2. It aids in splicing (i.e. the removal of introns from hnRNA)
3. It enhances translation of the mRNA
4. It regulates nuclear export (via the cap binding complex).

Polyadenylation

Although RNA polymerase continues to transcribe the DNA, the transcript is cleaved by the endonuclease activity of the polyadenylate polymerase complex approximately 10 to 30 nucleotides downstream from the polyadenylation signal (AAUAAA). The cleavage gives the 3′ end of the transcript a well-defined end. The 50–250 nucleotide poly(A) tail is generated from

Fig. 6.22 Initiation of transcription with eukaryotic RNA polymerase II. In the final stage, TFIIH phosphorylates amino acids in the tail of RNA Pol II, which reduces its affinity for TAFs and releases RNA Pol II for transcription. A, adenine; T, thymine; TAF, TATA-associated factor; TBP, TATA-binding protein – a saddle-shaped protein that unwinds the DNA helix; TF, transcription factor.

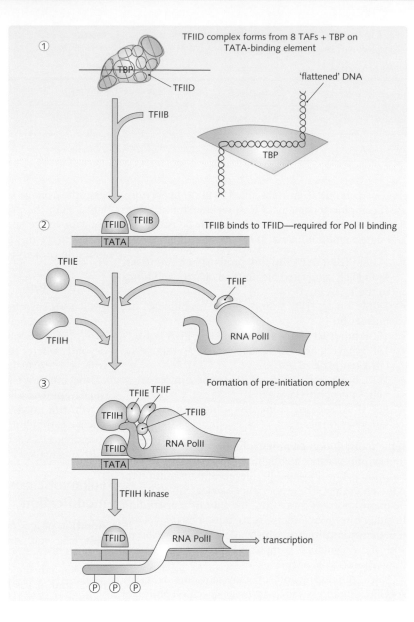

① TFIID complex forms from 8 TAFs + TBP on TATA-binding element

TBP

TFIID

'flattened' DNA

TBP

TFIIB

② TFIIB binds to TFIID—required for Pol II binding

TFIID TFIIB

TATA

TFIIE

TFIIF

TFIIH

RNA PolII

③ Formation of pre-initiation complex

TFIIE TFIIF

TFIIH

TFIIB

TFIID

RNA PolII

TATA

TFIIH kinase

TFIID

RNA PolII → transcription

P P P

adenosine triphosphate (ATP) by the polyadenylate polymerase complex (Fig. 6.24). The degree of polyadenylation correlates with the half life of the mRNA molecule; generally the longer the poly(A) tail, the more stable the mRNA and the longer its half life.

Splicing

HnRNAs are the primary transcripts from genomic DNA. The production of mature eukaryotic mRNAs from hnRNA involves a process called splicing. This is the removal of non-coding introns and the joining of intervening exons facilitated by a ribonucleoprotein complex called the spliceosome, which assembles immediately after the intron sequence has been transcribed. Spliceosomes consists of:

- a core structure of a number of subunits called small nuclear ribonuclear particles (snRNPs), consisting of RNAs and proteins. There are four major classes of snRNPs, named according to the snRNA they contain: U1, U2, U5 and U4/U6
- non-snRNP splicing factors
- an hnRNA.

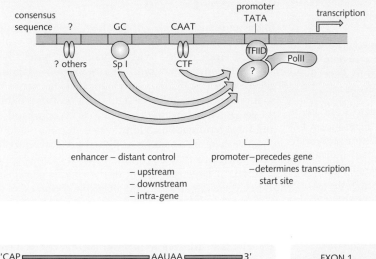

Fig. 6.23 Hierarchical control over gene expression in eukaryotes. Physiological levels of expression depend on the interaction between polymerase and transactivating and repressor proteins. There is sequence homology between TATAAT and eukaryotic TATA. Both form the site of the transcription initiation complex. A, adenine; T, thymine; C, cytosine; G, guanine.

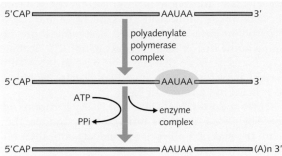

Fig. 6.24 Polyadenylation of mRNA. Cleavage occurs at the polyA site, which is usually found 10–35 nucleotides 3′ of the upstream polyA signal (AAUAAA). PolyA polymerase then adds several hundred A nucleotides to the 3′ end of the mRNA.

Clinical Note

Autoantibodies against snRNP proteins have been implicated in several autoimmune conditions. Components of the U1, U2 and U4–U6 snRNPs are antigenic targets in the anti-Smith antigen (Sm) response, present in about 30% of all patients with systemic lupus erythematosus. Autoantibodies against U1 snRNP-specific proteins (anti-nRNP) are specifically seen in mixed connective tissue disease. Both the Sm and nRNP antigens are required for the normal post-transcriptional, pre-messenger RNA processing to excise introns.

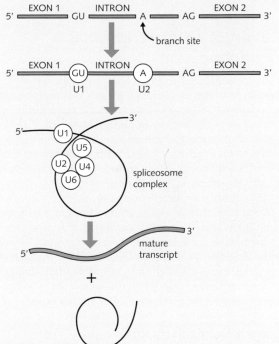

Fig. 6.25 Spliceosome mediated RNA splicing. The first stage in the reaction results in the breaking of the phosphate bond at the 5′ exon/intron boundary and its joining to the adenine at the branch site. In the second step the phosphate bond at the 3′ exon/intron boundary is cleaved, which is followed by the reformation of a phosphate bond between the terminal nucleotide of the first exon and the first nucleotide of the second exon.

The splicing process depends on the existence of consensus sequences within the hnRNA intron, which are recognized by components of the spliceosome (Fig. 6.25).

- The first two nucleotides of the intron are always GU, to which binds U1 forming the splice donor site.
- An 'A' nucleotide approximately 30 nucleotides from the 3′ end of the intron binds U2, forming the branch-point.
- The last two nucleotides are always AG, which binds U5 and forms the splice acceptor site.

It should be noted that RNA sequences contain many GU and AG sequences that are not used in splicing for poorly understood reasons.

The binding of the snRNPs to the intron causes it to form a loop. The splicing reaction then proceeds in two stages, releasing the first and second exons respectively (Fig. 6.25).

- The nucleotide at the branch site attacks the donor site and cleaves it and the 5′ end of the intron becomes covalently attached to the adenine nucleotide at the branch site, forming a 'lariat'-shaped structure.
- The 3′-OH end of the first exon (generated in the previous reaction) adds to the beginning of the second exon sequence, cleaving the RNA at the splice acceptor site.

After the reaction the exons are joined together and the intron sequence is released as a lariat.

Eukaryotic transcriptional regulation

Transcriptional regulation in eukaryotes is more complex than in prokaryotes. Although the DNA complement in all eukaryotic cell types is the same, the genes expressed from it vary. Differences in chromatin structure determine whether specific transcriptional activators and repressors are expressed. These bind to enhancers and silencers respectively and influence the levels of tissue-specific transcription of individual genes.

Even within active euchromatin, the expression of genes is regulated according to the needs of the organism. Thus, signalling cascades initiated by hormones in response to changes in the external environment will ultimately alter gene expression. Similarly, changes in the cellular internal environment may result in changes in gene expression (e.g. an accumulation of DNA damage will trigger the expression of apoptosis genes).

Epigenetic mechanisms

When a specific cell type replicates, the daughter cells retain the structural characteristics of the parent cell, suggesting that the changes in chromatin structure initiated in differentiation can be transmitted. Epigenetics is the exertion of a heritable influence on gene activity unaccompanied by DNA sequence change. Examples include histone modification and gene methylation (see p. 82).

EUKARYOTIC TRANSLATION AND PROTEIN SYNTHESIS

Translation is the mRNA-directed biosynthesis of polypeptides; a complex process involving several hundred macromolecules.

The genetic code

In the translation process from mRNA to protein, amino acids are coded for by groups of three bases called codons. Since nucleic acids contain four bases, there are 4^3 (64) possible codons. The same, non-overlapping, genetic code (Fig. 6.26) is seen in most living organisms, so it is considered universal. Out of 64 possible codons, the genetic code consists of 61 amino-acid coding codons and three termination codons, which stop translation.

Since there are 61 amino-acid coding codons, but only 20 amino acids that are commonly used in polypeptide synthesis, a large proportion of the code is considered to be degenerate, that is more than one codon exists for each amino acid. For example, the codons GGU, GGC, GGA and GGG all encode the amino acid glycine.

Codons in the mRNA transcript are recognized by the 3 nucleotide 'anticodon' tRNA molecules charged with the appropriate amino acid. The same tRNA may recognize codons differing in the third base, while those differing in the first or second bases are not. The 'wobble hypothesis' suggests that the third base in tRNA anticodons allow for a certain amount of play (or 'wobble'), so it may bind a variety of bases.

Three codons (UAA, UAG, and UGA) are not recognized by tRNAs, and these are termed stop codons. They mark the end of a polypeptide and signal to the ribosome to stop synthesis.

Mitochondrial DNA and the genetic code

Mitochondria contain their own unique DNA, which in humans consists of 16 kb of circular dsDNA coding for:

- 22 mitochondrial (mt) tRNAs
- two mt rRNAs
- 13 proteins synthesized by the mitochondrion's machinery; subunits of the oxidative phosphorylation pathway.

Figure 6.27 summarizes the differences between mitochondrial and nuclear DNA. Codon/anticodon pairings show more 'wobble' pairings than in the process originating in the nucleus due to unusual mt tRNA sequences such as mt tRNA^{ser}, which lacks a D arm.

Eukaryotic protein synthesis

Protein synthesis is very similar to that seen in prokaryotes but more associated factors are involved.

Initiation

The mechanism behind initiating translation in eukaryotes is poorly understood. There are two cap-dependent ways ribosomal subunits can reach the initiator AUG codon:

1. Scanning ribosome (Fig. 6.28)

Fig. 6.26 Standard genetic code. A, adenine; C, cytosine; G, guanine; T, thymine; U, uracil. To find out which amino acid a particular codon codes for, first select the 5′ end base (left column). Then read across to select the column corresponding to the second position base. Read down to find the 3′ end base (right column) and locate the row on which the corresponding amino acid lies. (Adapted from Nussbaum, McInnes and Willard, 2001.)

First base (5′)	Second base				Third base (3′)
	U	*C*	*A*	*G*	
U	UUU phe	UCU ser	UAU tyr	UGU cys	U
	UUC phe	UCC ser	UAC try	UGC cys	C
	UUA leu	UCA ser	UAA stop	UGA stop	A
	UUG leu	UCG ser	UAG stop	UGG trp	G
C	CUU leu	CCU pro	CAU his	CGU arg	U
	CUC leu	CCC pro	CAC his	CGC arg	C
	CUA leu	CCA pro	CAA gln	CGA arg	A
	CUG leu	CCG pro	CAG gln	CGG arg	G
A	AUU ile	ACU thr	AAU asn	AGU ser	U
	AUC ile	ACC thr	AAC asn	AGC ser	C
	AUA ile	ACA thr	AAA lys	AGA arg	A
	AUG met	ACG thr	AAG lys	AGG arg	G
A	GUU val	GCU ala	GAU asp	GGU gly	U
	GUC val	GCC ala	GAC asp	GGC gly	C
	GUA val	GCA ala	GAA glu	GGA gly	A
	GUG val	GCG ala	GAG glu	GGG gly	G

Fig. 6.27 Summary of variations between mitochondrial and standard genetic code. N, one of four nucleotides.

	Standard	Mammalian mitochondrion
UGA	Stop	Trp
AUA	Ile	Met (initiation signal)
CUN	Leu	–
AGA/AGG	Arg	Stop
CGG	Arg	–

2. An internal ribosome entry site (IRES) within the 5′-UTR (Fig. 6.28).

Most eukaryotic translation initiation is considered cap-dependent, involving ribosomal scanning of the 5′ untranslated region (5′-UTR) for an initiating AUG start codon. Both the proximity to the cap and the nucleotides surrounding the AUG start codon can influence the efficiency of the start site recognition during the scanning process. The scanning (40 S) ribosomal subunit will ignore a poor quality recognition site and skip potential starting AUGs, a phenomenon called leaky scanning. Therefore initiation is not always restricted to the AUG codon nearest the 5′ end. Like prokaryotes, the eukaryotic initiator tRNA carries a methionine residue. However, unlike prokaryotes, this methionine residue is not formylated.

Elongation

Like prokaryotes, the translation elongation cycle adds one amino acid at a time to a growing polypeptide according to the sequence of codons found in the mRNA, requiring eukaryotic elongation factors (eEFs).

Clinical Note

Diphtheria toxin is a bacterial exotoxin, encoded by a bacteriophage gene. It is secreted as a single polypeptide and cleaved into two fragments; one for toxin binding to the host cell membrane. The other contains enzymes (ADP-ribosylation) for the inhibition of elongation factor-2 (EF-2), inactivating the transfer of amino acids from tRNA to the polypeptide chain and inhibiting protein synthesis. Diphtheria toxin is very potent; a single molecule within a cell is lethal, inactivating millions of EF-2 molecules.

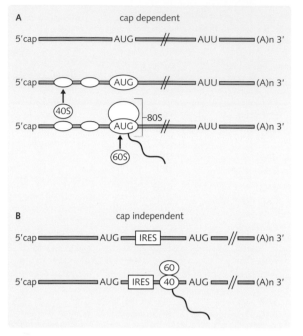

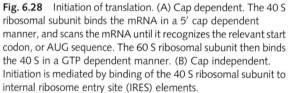

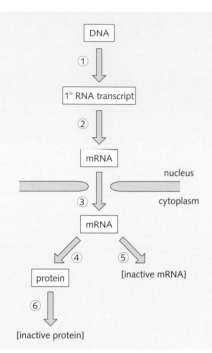

Fig. 6.28 Initiation of translation. (A) Cap dependent. The 40 S ribosomal subunit binds the mRNA in a 5′ cap dependent manner, and scans the mRNA until it recognizes the relevant start codon, or AUG sequence. The 60 S ribosomal subunit then binds the 40 S in a GTP dependent manner. (B) Cap independent. Initiation is mediated by binding of the 40 S ribosomal subunit to internal ribosome entry site (IRES) elements.

Fig. 6.29 Eukaryotic control of gene expression. (1) Transcription control. (2) Processing of transcript. (3) Transport control. (4) Translational control by selection of ribosomes by mRNA. (5) mRNA degradation control. (6) Protein activity control and post-translational modification.

Termination

All three stop codons, UAA, UGA and UGC, are recognized by the same specific eukaryotic releasing factor (eRF), which is composed of proteins eRF1 and eRF3. Binding of eRF to a stop codon releases mRNA from the ribosome, which then dissociates into its constituent subunits ready to reassemble on another molecule for a new round of protein synthesis.

CONTROL OF GENE EXPRESSION AND PROTEIN SYNTHESIS

Control of protein synthesis enables control of gene expression. In prokaryotes, control can be at transcription or translation, whereas in eukaryotes a total of six control points have been identified (Fig. 6.29).

Constitutive, inducible and repressible enzymes

Constitutive enzymes exist at fixed concentrations in the cell, irrespective of changes in its environment, one of the products of housekeeping genes in multicellular organisms.

The level of expression of inducible/repressible enzymes is altered by the chemical composition of the cellular environment.

Clinical Note

Many hepatic drug-metabolizing enzymes are induced or inhibited by therapeutic drugs and other compounds. The P450 enzyme subfamily CPY3A is the most abundant of the hepatic cytochrome enzymes. It can be induced (e.g. by phenytoin and rifampicin) or inhibited (e.g. by erythromycin and fluoxetine). Induction of these enzymes can accelerate drug breakdown and reduce the therapeutic window, while inhibition prolongs drug activity and is potentially toxic.

POST-TRANSLATIONAL MODIFICATION OF PROTEINS

Concepts

Post-translational modification (i.e. modifications after translation) give mature proteins functional activity. Modifications include peptide cleavage and covalent modifications, such as glycosylation, phosphorylation, carboxylation and hydroxylation of specific residues.

A newly synthesized protein may be destined for extracellular secretion, the cytoplasm or organelles, such as the plasma membrane, nucleus or lysosomes. Proteins are directed to the appropriate location by:

- conserved amino-acid sequence motifs (e.g. the signal peptide, the nuclear targeting signal)
- moieties added by post-translational modification (e.g. mannose-6-phosphate for lysosomal delivery).

Signal peptide

The signal peptide (or leader sequence) is a characteristic hydrophobic amino acid sequence of 18–30 amino acid residues near the amino terminus of the polypeptide that directs non-cytoplasmic polypeptides into the ER lumen during translation (Fig. 6.30).

Inside the ER lumen the polypeptide can undergo post-translational modifications specific to each protein. Proteins destined for extracellular secretion are transported to the Golgi apparatus in vesicles that bud off the ER. Further modification may take place here and the secreted protein is packaged into vesicles that

fuse with the cell plasma membrane to release their contents to the exterior.

Glycosylation of proteins

Glycosylation occurs in the ER lumen or the Golgi apparatus, involving the addition of an oligosaccharide to specific amino-acid residues. There are two types of glycosylation, designated N-linked and O-linked, that employ specific glycosyltransferases:

- N-linked – oligosaccharide added to the polypeptide by a β-N-glycosidic bond to an aspartate residue
- O-linked – oligosaccharide joined by α-O-glycosidic bond to a serine or threonine residue.

Clinical Note

I-cell disease (mucolipidosis type II) is a lysosomal storage disease resulting from a deficiency of mannose-6-phosphate glycosyltransferase, which is responsible for glycosylation of enzymes destined for lysosomes. As the cell's lysosomal enzymes lack their lysosomal uptake recognition markers, they are secreted into the extracellular matrix, and

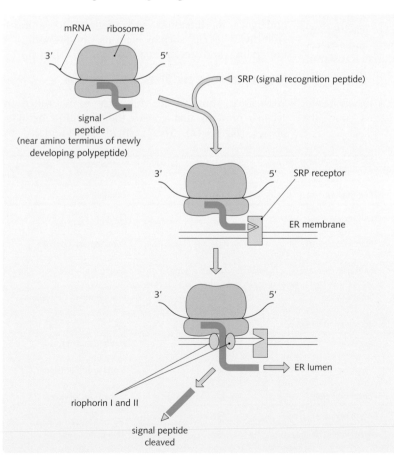

Fig. 6.30 Role of the signal peptide in translocation into the endoplasmic reticulum (ER) lumen. The signal peptide is near the amino terminus of the newly developing polypeptide. It associates with a cytoplasmic signal recognition peptide (SRP) and then with an SRP receptor ('docking protein') on the ER membrane. The ribosome then interlocks between two membrane-associated proteins, riophorin I and II, which drive the developing polypeptide into the ER lumen.

undigested substrates accumulate within the lysosomes with severe pathological consequences, including rapidly progressive growth failure, severe developmental delay, skeletal deformation and early death.

Glycosylation is important for either protein function or compartmentalization:

- O-linked glycosylation – production of blood group antigens
- N-linked glycosylation – transfer of acid hydrolase to lysosomes; production of mature antibodies.

Other modifications of protein

Proteins may be modified by (Fig. 6.31):

- phosphorylation – targeting Ser, Thr or Tyr residues and regulating enzyme (Ser) or protein activity (Tyr). Kinases transfer phosphate groups from ATP onto the target residue
- sulphation – targeting Tyr, important in compartmentalization (e.g. marking proteins for export) and biological activity
- hydroxylation – targeting Lys and Pro residues, important in the production of collagen (and extracellular matrix protein). This occurs during translation and is essential for the formation of the collagen triple helix
- lipidation of Cys and Gly residues – for anchoring proteins, such as antibody receptors, into the membrane

- acetylation of Lys – changing the charge of the residue or its binding properties to DNA
- cleavage – activating some enzymes and hormones.

CELL CYCLE

Concept of the cell cycle

The cell cycle is a controlled set of events, culminating in cell growth and division into two daughter cells. These events are ordered into pathways in which the initiation of late events, such as cell division, is dependent on the successful completion of early events, such as DNA synthesis. The cell cycle consists of four phases (Fig. 6.32):

1. G_1 – the gap between mitosis of the preceding cell cycle and DNA synthesis of the current cycle. It contains the restriction point; the start of the cell cycle
2. S – the DNA synthesis phase; the cell's chromosomes are replicated
3. G_2 – the gap between DNA synthesis completion and the decision to divide
4. M/mitosis – the cell division phase resulting in the production of two daughter cells.

Together, G_1, S and G_2 are known as interphase, the interval between divisions during which the cell undergoes its functions and prepares for mitosis. Non-dividing cells, such as neurons, are quiescent, i.e. not cycling, and remain in a resting state called G_0.

Regulation of the cell cycle

Progression through the cell cycle is controlled by cyclins, proteins that govern the transition between phases (Fig. 6.32). Their activity is modulated by

Fig. 6.31 Summary of some post-translational modifications.

Destination	Protein function	Modification	Residue	Example
Secreted	Structural	Hydroxylation	Lys/Pro	Collagen
	Enzyme	Hydrolytic cleavage	(peptide bond)	Pepsinogen → pepsin
	Hormone	Hydrolytic cleavage	(peptide bond)	Proinsulin → insulin
	Clotting factor	Carboxylation	Glu	Prothrombin → thrombin
	Antibody	Glycosylation (N-linked)	Asp	IgG
Membrane	Receptor	Lipidation	Gly/Cys	Antibody receptors
	Cell recognition	Glycosylation (O-linked)	Ser	Blood group antigens
	Receptor activation	Phosphorylation	Tyr	Growth factor receptor
Cytoplasm	Enzyme	Phosphorylation	Tyr/Ser	
Lysosome	Hydrolytic enzymes	Glycosylation (N-linked)	Asp	Acid hydrolases

Fig. 6.32 The cell cycle – a summary.

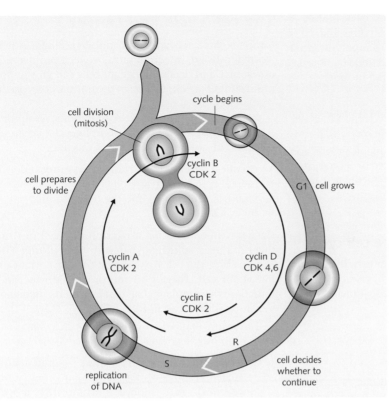

changes in their intracellular concentrations, which is regulated by both mRNA expression level and the protein level. In the presence of DNA damage and structural problems 'checkpoint proteins' stall the cell cycle, allowing DNA repair to occur, preventing mutation propagation.

Cyclin-dependent kinases

Forward progression though the cell cycle is controlled by protein kinases, the activity of which varies at different cycle phases. These are noncovalent complexes of an activating protein called a cyclin, and a catalytic subunit called a cyclin-dependent kinase (CDK). Activated CDKs are serine/threonine kinases which stimulate cell-cycle progression by phosphorylating specific proteins in the cell required for transition. For example:

- at the beginning of prophase of mitosis, nuclear membrane breakdown is initiated by lamin phosphorylation to form part of the nuclear skeleton
- chromosome condensation at the beginning of mitosis is initiated by the phosphorylation of H1 histone, a nuclear-associated protein
- G_1–S transition is initiated by CDK-dependent phosphorylation of Rb protein. Unphosphorylated Rb

protein forms a complex with the transcription factor E2F. On phosphorylation of Rb, E2F is released and activates transcription of genes required for the transition from G_1 into S phase.

Human cyclins are typically designated A, B, D and E. Each one accumulates at a different time in the cell cycle (Fig. 6.33).

Maturation-promoting factor (MPF) is an example of a cyclin–CDK complex. It initiates transition from G_2–M, under the control of cyclin B. A small increase in cyclin levels produces a big increase in MPF kinase activity, promoting chromosomal condensation.

Fig. 6.33 Cyclins and cyclin-dependent kinases (CDKs) involved in the cell cycle. Cyclin C has recently been discovered and, along with CDK8, is thought to regulate RNA transcription during the cell cycle.

Cyclin	Kinase	Function
D	CDK4, CDK6	Progression past restriction point at G_1/S boundary
E, A	CDK2	Initiation of DNA synthesis in early S phase
B	CDK1	Transition from G_2 to M

CDKs are negatively controlled by CDK inhibitors (CKIs). When cyclin levels rise above a threshold, CKIs cannot exert their effect. They are also temporarily arrest the cell cycle due to DNA damage and unfavourable environmental conditions. CDKI families include:

- INK4 – inhibits CDK4 and CDK6
- CIP/KIP – inhibits G1/S CDKs and S phase CDKs, activates cyclin D-CDK4.

The activity of CDKs and subsequent progression through the cell cycle, is also influenced by several extracellular signalling pathways. These facilitate coordinated cell division in multicellular organisms.

Extracellular regulation of the cell cycle

A mitogen is an agent that induces mitosis. In addition to cyclins, the cell cycle is influenced by several factors that activate the mitogen activated protein kinase (MAPK) pathways, including:

- growth factors
- hormones
- cell–cell interactions.

The huge variety of factors involved allows fine control of cell growth and replication, and response to environmental changes. Growth factors are soluble substances that can act locally or remotely to affect cell growth. Growth factors, like hormones and cell–cell interactions, act by binding to specific cell surface target receptors. This initiates the phosphorylation of target proteins within the cell in a cascade that alters gene expression.

Checkpoints

There are numerous points of control in the cell cycle where progression through it may be regulated, including:

- the restriction point during G_1 – here the cell becomes independent of external mitogenic stimuli and becomes committed to completing a cycle. The cell will not proceed if there are inadequate nutrients or growth factors available
- the G_1–S DNA integrity checkpoint ensures the previous cycle of division is complete and any damage repaired before DNA synthesis. This is the main site of p53 action (see p. 97)
- the G_2–M DNA integrity checkpoint ensures DNA synthesis, and that any DNA damage has been repaired before mitosis
- the spindle-assembly checkpoint ensures the prerequisites for chromosome segregation have been met before chromosome segregation.

The cell cycle and cancer

Cancer is characterized by uncontrolled cellular growth. Normal cells are in equilibrium between proliferation, quiescence and death. Malignant cells can grow autonomously and are not subject to the normal controls that regulate cell division.

Malignancy is due to DNA mutations resulting in increased or decreased expression of genes of cell cycle control. The commonest causes of this are:

- chemical damage (e.g. by benzene, nitrosamines)
- radiation (e.g. ultraviolet light)
- viral DNA integrated into the host genome
- inherited defects.

Cancer represents clonal expansion of a cell in which there has been sufficient change to the genomic DNA to transform the cell's phenotype from a normal to a malignant cell. Usually in the progression to cancer mutations accumulate that together cause malignant transformation (see Ch. 8).

The genes that, when mutated, are associated with cancer can be categorized as:

1. Oncogenes (e.g. *Ras Fos Myc*) – mutated/up-regulated versions of normal cellular genes (proto-oncogenes) that induce uncontrolled growth. Proto-oncogenes frequently make up cell signalling pathways (see Ch. 3)
2. Tumour suppressor genes (e.g. *p53 Rb GAP*) – expressed in normal cells, with loss of activity resulting in uninhibited growth. Many code for proteins that are normally involved in regulating cell division and differentiation
3. Apoptotic regulatory genes (e.g. *BCL2*, caspases) – expressed in normal cells to control the apoptotic process. Inappropriately activated anti-apoptotic genes can drive cancer, as can the loss of pro-apoptotic genes
4. DNA repair genes (e.g. *XPA BRCA1*) – for genome integrity. Inability to repair DNA damage allows mutations to be passed on to subsequent cell generations, and immortalization. Tumour cells commonly demonstrate genomic instability.

The NF1 gene is an example of a tumour suppressor gene encoding a Ras-GAP. It is mutated in neurofibromatosis, meaning that Ras is more likely to be activated, resulting in uncontrolled growth.

> ### HINTS AND TIPS
>
> Tumour suppressor genes are responsible for stalling the cell cycle if there is DNA damage, in order to allow time for repair. They are like the brakes on a car.

Proto-oncogenes are generally responsible for telling a cell it may divide. When mutated or deranged, they inappropriately promote cell division and survival. They are like the accelerator.

For cancer to arise you need to remove the brake (i.e. mutate a tumour suppressor gene, such as *p53*), and then apply the accelerator (i.e. turn on/up a mitogen or an anti-apoptotic gene, or turn off/down a pro-apoptotic one).

p53

p53 is a tumour suppressor dubbed the 'guardian of the genome'. Its basic function is to restrict entry of cells with damaged DNA into S phase (i.e. regulates progression past the restriction point). Cells with mutant p53 are not arrested in G_1 and progress through the cell cycle and division with damaged DNA. The *p53* gene lies on chromosome 17 and codes for a nuclear phosphoprotein of 53 kDa, having three major roles:

1. Transcription activator – regulating certain genes involved in cell division
2. G_1 checkpoint for DNA damage – in excess DNA damage (e.g. ultraviolet damage) it arrests cell division, allowing repair
3. Participation in initiating apoptosis

Mechanisms of apoptosis

Apoptosis comprises three different mechanisms of programmed cell death:

1. Internal signals – the Bcl-2 protein displayed on outer mitochondrial membranes inhibits apoptosis. Bcl-2 is inhibited when internal damage occurs as the Bax protein punches holes in this outer mitochondrial membrane. This action of Bax activates caspases that ultimately leads to damaged-cell phagocytosis
2. External signals – complementary death activators bind the Fas and TNF receptor domains at the cell surface. This transmits a cytoplasmic signal that activates caspases, initiating a caspase cascade culminating in the phagocytosis of the marked cell
3. Apoptosis-inducing factor (AIF) – some cell types, such as neurons, undergo cell death not mediated by caspases. The AIF protein usually resides in the mitochondrial intermembrane space. However, should the cell receive a signal to induce apoptosis, AIF is released from the mitochondria and binds to nucleic DNA. This leads to DNA destruction and eventually, cell death.

MITOSIS AND MEIOSIS

Overview of cell division

Cell division is the process by which a cell, including the nucleus, replicates and splits to produce two daughter cells. Many somatic cell types are continually replenished by cell division. In order to be viable, each daughter cell must contain a complete set of genetic material so that all proteins can be expressed at appropriate levels.

In addition to its role in directing protein synthesis, DNA enables the passage of genetic information from one generation to the next. Therefore, in the sexually reproducing multicellular organism, cells must have two mechanisms of cell division, resulting in both diploid and haploid daughter cells.

- Mitosis is cell division in somatic cells resulting in two genetically identical daughter cells.
- Meiosis occurs in gamete formation (e.g. sperm and ova). Each daughter cell contains half the genetic information of the parent cell and crossing-over ensures reassortment of genetic material between homologous ('paired') chromosomes.

Mitosis

Two genetically identical diploid daughter cells are produced.

$$\text{Mitosis} : 2n \rightarrow 2n$$

There are six distinct phases (Fig. 6.34).

Prophase

The cell's chromatin condenses into the classic chromosome structure with each duplicated chromosome identifiable as a pair of sister chromatids joined by the duplicated, but unseparated, centromere. The centrioles duplicate and migrate towards opposite poles of the cell. A spindle of microtubules is formed simultaneously. The nucleoli disperse.

Prometaphase

This is the breakdown of the nuclear envelope and formation of kinetochores, points of attachment between the chromosome and the spindle, at the centromeres of the chromosomes.

Metaphase

The chromosomes become attached to the spindle at the kinetochore and are arranged along it, forming the equatorial plate. At this point the chromosomes are maximally condensed and are most visible.

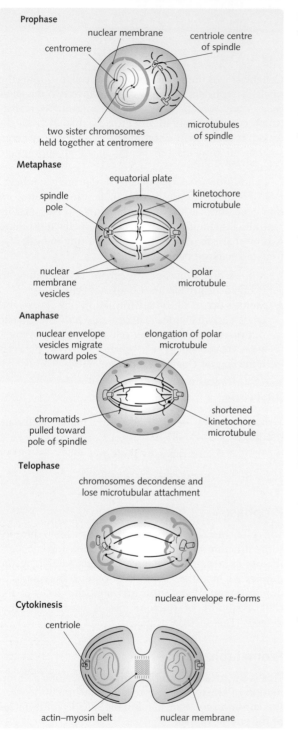

Prophase

nuclear membrane

centromere

centriole centre of spindle

two sister chromosomes held together at centromere

microtubules of spindle

Metaphase

equatorial plate

spindle pole

kinetochore microtubule

nuclear membrane vesicles

polar microtubule

Anaphase

nuclear envelope vesicles migrate toward poles

elongation of polar microtubule

chromatids pulled toward pole of spindle

shortened kinetochore microtubule

Telophase

chromosomes decondense and lose microtubular attachment

nuclear envelope re-forms

Cytokinesis

centriole

actin–myosin belt

nuclear membrane

Fig. 6.34 Mitosis. (Adapted from Stevens and Lowe, 1997.)

Anaphase

The centromeres separate allowing the chromatids to be pulled to opposite poles by the spindle. At the end of anaphase a complete set of chromosomes clusters at each pole of the cell.

Telophase

The chromosomes uncoil, assuming the characteristic extended state. The nuclear membrane re-forms and nucleoli reappear.

Cytokinesis

A cleavage furrow forms around the mid-region between the poles, dividing the cytoplasm into two and ultimately leading to the formation of two daughter cells each with a complete diploid chromosome complement.

> **HINTS AND TIPS**
>
> Each sister chromatid of a metaphase chromosome is a double helix, because both strands were replicated in the preceding S phase. Thus, after separation in mitotic anaphase each daughter cell receives a complete copy of the genome. (It is a common misconception that the strands of the double helix are separated at anaphase, in which case the genome would only be complete after S phase.)

Meiosis

In the first division of meiosis two genetically different haploid cells are formed (Fig. 6.35). In the second division, each haploid cell is duplicated.

$$\text{Meiosis} \to 2n \to n$$

Prophase I

There are five stages during which homologous chromosomes come together and exchange segments in homologous recombination:

1. Leptotene – spindle forms
2. Zygotene – homologous chromosomes pair, shorten and thicken, and form bivalents (pairs of homologous chromosomes)
3. Pachytene – chiasmata begin to form; points at which non-homologous chromatids become associated with each other via base pairing for 'crossing-over' between the chromatids
4. Diplotene – exchange of genetic material in chiasmata and nuclear membrane disappears
5. Diakinesis – recombinant chromosomes are formed.

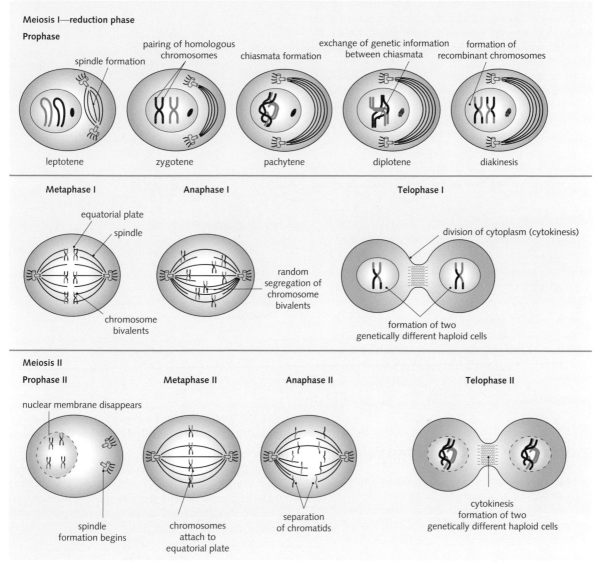

Fig. 6.35 Meiosis – a summary.

Metaphase I

Chromosomes become attached to a spindle (see mitotic metaphase).

Anaphase I

The chromatids do not separate and whole chromosomes migrate to opposite poles of the spindle, hence 'reduction division'.

Telophase I

Two genetically different haploid cells are formed.

Second division

The second division is like mitosis, but with a haploid number of chromosomes. The chromatids separate in anaphase II.

Genetic diversity and gametogenesis

Two processes in meiosis are vital to generate genetic diversity:

1. Chiasmata formation ('crossing-over'), which allows random exchange of genetic material between homologous chromosomes
2. Independent segregation of homologous chromosomes.

During anaphase I, homologous chromosomes segregate independently of each other. Since humans possess 23 pairs of homologous chromosomes, there are 2^{23} possible ways that the chromosomes can segregate to form a haploid set.

In humans, meiosis begins during gametogenesis. In females this occurs in the ovaries. The first division begins during the 5th month of embryonic life, but it is arrested at pro-metaphase and completed just before ovulation. Meiosis II takes place after ovulation. Therefore:

- there is a fixed number of oocytes
- there is a period of arrest between the start of meiosis I and the completion of meiosis II of 12–45 years.

Clinical Note

In oogenesis, meiosis is arrested in pro-metaphase of meiosis I, with a long latent period (dictyotene phase). This lengthy interval may lead to an accumulation of wear and tear effects on the primary oocyte, damaging the cell's spindle formation and repair mechanisms, and predisposing to a failure of chromosome separation at meiosis I and chromatid separation at meiosis II (non-disjunction). Non-disjunction at meiosis I has been implicated in trisomies such as Down syndrome (trisomy 21), Edward syndrome (trisomy 18) and Patau syndrome (trisomy 13).

Oogenesis produces a single oocyte and two polar bodies (Fig. 6.36).

Male spermatogenesis occurs in the seminiferous tubules of the testes. After sexual maturity the spermatogonia continuously multiply by mitosis, subsequently undergoing meiosis to produce unlimited numbers of spermatocytes (Fig. 6.37).

Endoreduplication

Endoreduplication, or endomitosis, is repeated DNA replication in the absence of nuclear division and cytokinesis (Fig. 6.38). It can generate huge nuclei with up to 16 copies of the DNA. Endoreduplication occurs in the formation of megakaryocytes, which are cells in the bone marrow from which anucleate platelets bud off.

DNA DAMAGE AND REPAIR

DNA damage

DNA damage is structural change to the DNA molecule, which interferes with replication and transcription, and sequence changes, which may disrupt base pairs or lead to the incorporation of an incorrect base into the

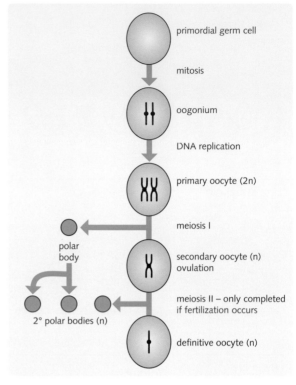

Fig. 6.36 Oogenesis.

replicating DNA strand. Two main types of damage occurs at a rate of 1000–1 000 000 molecular lesions per cell per day:

1. Endogenous – caused by the products of normal metabolism, from errors in DNA replication and spontaneously arising such as spontaneous deamination of adenine and cytosine producing hypoxanthine and uracil residues
2. Exogenous – caused by external agents.

Mutagens

Agents that cause exogenous DNA damage are known as mutagens, including:

- ionizing radiation, such as γ-rays and X-rays
- ultraviolet light – promotes chemical cross-linking between two adjacent thymine residues on a DNA strand, resulting in a pyrimidine dimer, distorting the DNA double helix in this region
- chemical mutagens, which can be of three types:
  - base analogues (e.g. 5-bromouracil; resembles thymine and pairs with adenine) – incorporate into DNA and cause misreading
  - chemical modifiers (e.g. hydroxylamine and compounds containing and propagating free radicals, such as those formed during the metabolism of polycyclic aromatic hydrocarbons in

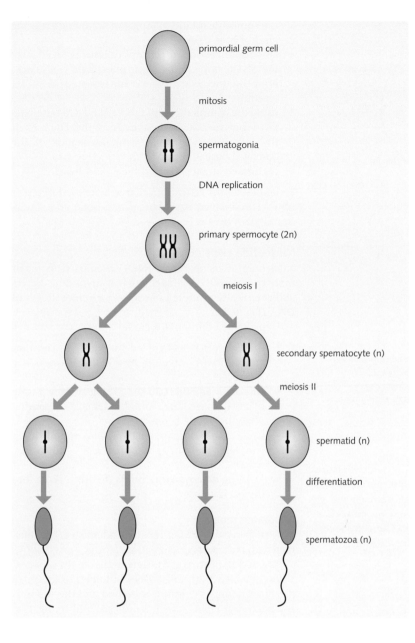

Fig. 6.37 Spermatogenesis.

tobacco smoke) – react with bases to form derivatives that cause misreading
- intercalators (e.g. some antibiotics and heavy metals) – slip between adjacent bases and inhibit RNA transcription
- viral genomes – can become incorporated into eukaryotic chromatin, disrupting coding or promoting regions, or it can affect levels of expression of existing genes.

Mutations

A mutation is a change in the base sequence of DNA (see Ch. 8).

- A change in the number of bases in a coding region, for instance due to insertion or deletion, it may result in a 'frameshift' error.
- An altered base or point mutation may cause a misreading error, resulting in an altered protein product.
- Double-strand DNA breaks may result in a chromosomal level rearrangement, such as a translocation.
- The mutation may disrupt a regulatory region and affect the level of expression of a particular gene.

DNA damage may be repaired or if extensive, trigger apoptosis.

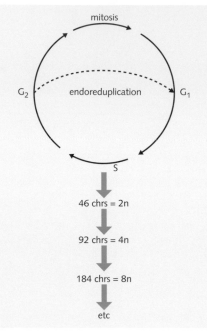

Fig. 6.38 Endoreduplication. Chromosomal duplications without intervening mitoses, resulting in increases in DNA content and cell enlargement. Megakaryocytes undergo endoreduplication, with an average ploidy of 16n (range 4n–64n).

DNA repair

DNA repair mechanisms act as a major defence against damage to DNA and minimize cell-killing, mutations, replication errors, persistence of DNA damage and genomic instability. Abnormalities in these processes have been implicated in cancer and ageing. There are several systems for the detection and repair of various types of DNA damage (see Fig. 6.27).

Direct reversal of DNA damage

One of the most frequent causes of point mutation in the human genome is the alkylation (especially methylation) of bases at specific sites. If unrepaired, such lesions can lead to the incorporation of incorrect bases in subsequent replication. Such damage is reversed by 'suicide' enzymes known as the DNA alkyltransferases, which are consumed by the process of de-alkylation.

Single-stranded damage

Base excision repair
Base excision repair (BER) is a process by which a damaged or inappropriate base is removed from its sugar linkage and replaced (see Fig. 6.27) and involves:

- removal of the damaged base by a glycosylase, leaving an abasic or AP site. A specific DNA glycosylase

is responsible for identifying and removing a specific kind of base damage

- an endonuclease and phosphodiesterase recognize the AP site and cut the sugar phosphate DNA backbone on the 5′ of the AP site, creating a 3′-OH terminus
- using the complementary DNA strand as a template, DNA polymerase β then extends the DNA from the free 3′-OH to replace the nucleotide of the damaged base
- the nicked strand is sealed by a DNA ligase.

Variations of this general mechanism are used to repair up to 10 damaged bases, sugar backbone and single-strand DNA breaks.

Nucleotide excision repair
Nucleotide excision repair (NER) is utilized to repair more complex DNA damage, such as thymidine dimers and chemically modified bases over a longer stretch of DNA (Fig. 6.39).

Steps involved in NER include:

- the damage is recognized by a protein complex and the DNA unwound by a helicase to produce a 'bubble'
- cuts are made on both the 3′ side and the 5′ side of the damaged area and the damage-containing oligonucleotide is removed
- using the complementary DNA strand as a template, the gap is filled by DNA polymerase δ and ε
- DNA ligase covalently binds the fresh piece into the backbone.

Mismatch repair
Mismatch repair (MMR) repairs undamaged, but mismatched, base pairs and small insertions or deletions.

The overall process of MMR is similar to the other excision repair pathways. The DNA lesion is recognized, a patch containing the lesion is excised and the strand is corrected by DNA repair synthesis and re-ligation (Fig. 6.39).

Double-stranded damage

Homologous recombination repair
Homologous recombination repair (HRR) typically occurs between DNA sequences with extended homology. On detection of a DNA double-stranded break (DSB), each of the 5′ ends of the break are resected by exonucleases, to leave long sections of 3′-ended single-stranded DNA (ssDNA) tails. These 3′ ends are recombinogenic and can invade a suitable homologous template molecule (such as the sister chromatid or homologous chromosome). The invading ssDNA ends then act as primers for DNA synthesis using the invaded

Fig. 6.39 Mammalian DNA defect repair mechanisms. Guanine, adenine, thymine, and cytosine (GATC) exonuclease recognizes and binds to a GATC base sequence near to the DNA defect and nicks the DNA at this site.

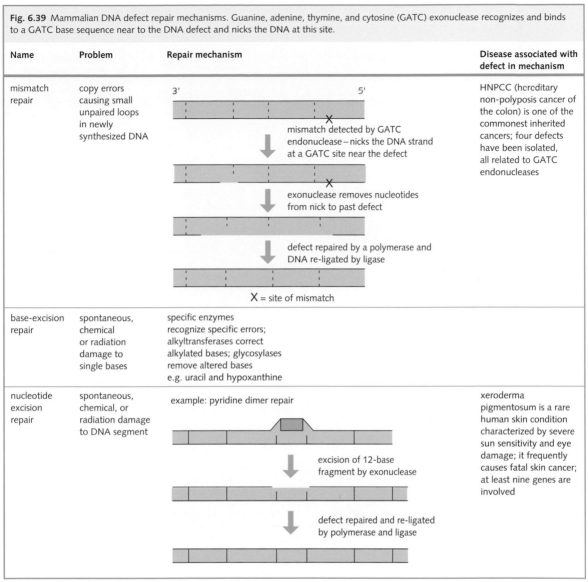

Name	Problem	Repair mechanism	Disease associated with defect in mechanism
mismatch repair	copy errors causing small unpaired loops in newly synthesized DNA	*[diagram]* mismatch detected by GATC endonuclease – nicks the DNA strand at a GATC site near the defect; exonuclease removes nucleotides from nick to past defect; defect repaired by a polymerase and DNA re-ligated by ligase. X = site of mismatch	HNPCC (hereditary non-polyposis cancer of the colon) is one of the commonest inherited cancers; four defects have been isolated, all related to GATC endonucleases
base-excision repair	spontaneous, chemical or radiation damage to single bases	specific enzymes recognize specific errors; alkyltransferases correct alkylated bases; glycosylases remove altered bases e.g. uracil and hypoxanthine	
nucleotide excision repair	spontaneous, chemical, or radiation damage to DNA segment	example: pyridine dimer repair *[diagram]* excision of 12-base fragment by exonuclease; defect repaired and re-ligated by polymerase and ligase	xeroderma pigmentosum is a rare human skin condition characterized by severe sun sensitivity and eye damage; it frequently causes fatal skin cancer; at least nine genes are involved

Continued

homologous molecule as a template. Thus, synthesis results in the restoration of the degraded single-strands (Fig. 6.39).

Non-homologous end joining

Non-homologous end joining (NHEJ) is regarded as an illegitimate repair pathway. Unlike HRR, NHEJ does not require an undamaged partner molecule and does not rely on extensive homologies between the two recombining ends. In most cases, direct end-ligation of a DSB is not possible, due to the presence of damaged bases and sugar moieties flanking the DSB and DNA. DSBs will require some processing before they can be ligated. This means NHEJ is rarely error-free and various sequence deletions are usually introduced around the area of the original DSB (Fig. 6.39).

Fig. 6.39 Mammalian DNA defect repair mechanisms. Guanine, adenine, thymine, and cytosine (GATC) exonuclease recognizes and binds to a GATC base sequence near to the DNA defect and nicks the DNA at this site—cont'd

Name	Problem	Repair mechanism	Disease associated with defect in mechanism
homologous recombination repair	replication induced + replication independent double strand breaks (spontaneous, chemical or radiation induced)		polymorphisms in HRR genes have been implicated in AML
non-homologous end joining	replication induced + replication independent double-strand breaks (DSB) (spontaneous, chemical or radiation induced)		severe combined immunodeficiency syndrome (SCID) is an inherited disorder characterized by decreased T and B cell function. Opportunistic infections occur readily, leading to death

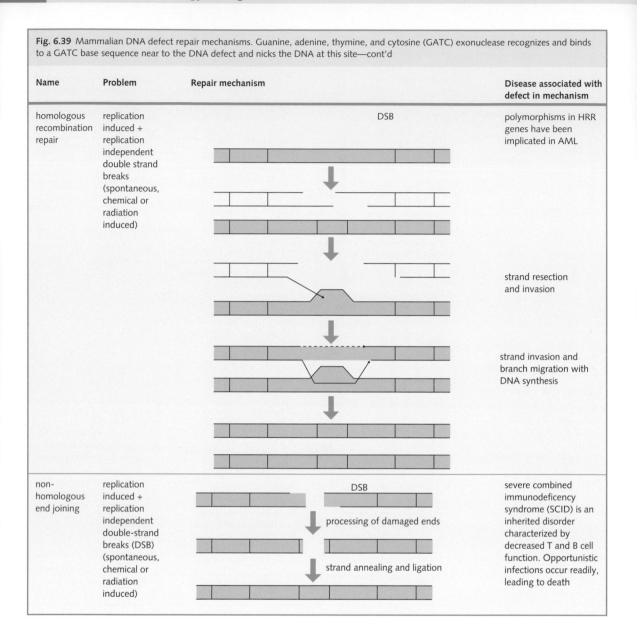

DSB

strand resection and invasion

strand invasion and branch migration with DNA synthesis

DSB

processing of damaged ends

strand annealing and ligation

Tools in molecular medicine 7

● Objectives

By the end of this chapter you should be able to:
- Understand the process of nucleic acid hybridization.
- Explain important DNA amplification techniques including cloning and PCR.
- Compare and contrast Southern, northern and western blotting techniques.
- Understand the differences between polymorphic markers including single nucleotide polymorphisms and variable number tandem repeats.
- Describe three different cytogenetic methods for examining chromosome structure.
- Describe the main aims of the human genome project and appreciate the important role of bioinformatics in the postgenomic era.
- Understand the principles of genetic linkage and its role in identifying human disease genes.
- Define the terms genome, transcriptome, proteome and metabolome.
- Describe some of the important treatments that have been developed with the use of genetic engineering.

MOLECULAR TECHNIQUES

Our understanding of the molecular basis of disease follows the discovery of the structure of DNA and the development of new technologies permitting detailed analysis of genes. In addition to providing the basis for theoretical advances, the techniques of molecular genetics permit the reliable detection and diagnosis of genetic disease. It is a fast-moving field in which new techniques are regularly introduced.

HINTS AND TIPS

The National Center for Biotechnology Information (http://www.ncbi.nlm.nih.gov) is a gateway into a range of public databases, including scientific literature (e.g. PubMed and Online Mendelian Inheritance in Man (OMIM) – a catalogue of human single gene defects) as well as molecular databases and access to bioinformatics tools.

Molecular genetics is the study of the structure and function of genes at the molecular level. Techniques concern:
- separation of nucleic acids from the other components of the cell
- characterization of DNA sequences
- study of gene expression
- manipulation and modification of DNA
- gene cloning and mapping.

HINTS AND TIPS

Molecular geneticists face two problems:
1. Obtaining sufficient quantities of nucleic acid to work with
2. Identifying specific sequences within a complex mixture of sequences.

Most of the techniques of molecular genetics address one or both of these.

The issue of complexity

A central issue within molecular analysis is that whether looking at DNA, RNA or protein, the molecules of interest are very often obscured by the large number of other molecules of similar type. In order to 'see the wood through the trees', molecular geneticists use a number of different techniques to identify and analyse the targets of interest, and these will be explored in this chapter.

Preparation

Before a target molecule can be isolated and analysed, it must first be obtained from the mixture in which it resides (often the contents of a cell). This is followed by one or more of the following.

- Separation – reducing the complexity of the mixture (e.g. by electrophoresis or chromatography).

- Detection – interrogation of the mixture for the presence of the target (e.g. using hybridization or reaction with specific antibodies).
- Isolation – purification (e.g. by cloning or a series of separation steps).
- Amplification – to obtain sufficient amounts of the molecule to work with (e.g. by cloning or PCR).

Nucleic acid hybridization

Every technique used to detect specific sequences of DNA or RNA relies on hybridization (Fig. 7.1).

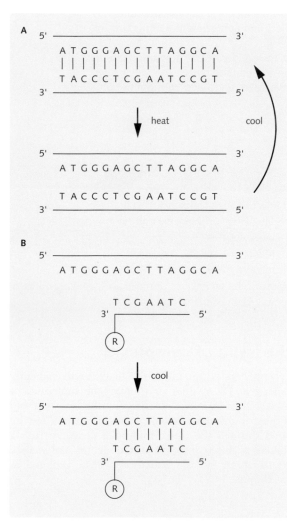

Fig. 7.1 DNA denaturation and hybridization. Hydrogen bonds between paired bases are broken, leading to separation of the DNA strands (denaturation). These may form new base pairs with any complimentary single stranded nucleic acid (hybridization). (A) Re-annealing of the two complimentary strands. (B) Hybridization of denatured DNA with a fluorescently labelled (R) probe. In practice, the probe would be ~20bp long.

Heating double-stranded DNA separates the two strands by breaking the hydrogen bonds holding the base pairs together; this is called 'denaturation' or 'melting'. Resulting single strands can re-anneal to re-form the original double stranded DNA. More importantly, single strands (DNA and/or RNA) that were not originally part of the same molecule can pair with each other; termed 'hybridization'. Stable hybrids are formed only when the sequences of the two strands are complementary. Hybridization allows the detection of specific sequences in complex mixtures; for example a radio-labelled DNA 'probe' can be used to specifically detect a particular sequence that occurs only once in the entire genome; similarly the specificity of amplification of PCR results from hybridization of primers with the target sequence.

Separation and detection

Gel electrophoresis

Gel electrophoresis is used to isolate molecules based on their molecular size or charge, from mixtures of DNA, RNA or protein. Mixtures are loaded into wells of a gel commonly made from agarose (DNA or RNA electrophoresis) or polyacrylamide (DNA sequencing or protein electrophoresis). An electric field is applied to the gel, which causes:

- charged molecules to migrate through the electric field (e.g. negatively charged nucleic acids migrate towards the anode)
- small molecules migrate faster than large and end up furthest from the wells
- sizes of molecules can be estimated by reference to 'marker' molecules of known sizes.

Once the molecules have been separated they must be visualized in some way.

Staining
Stains which bind to any nucleic acid or protein are used to visualize the separated molecules on gels. DNA and RNA products can be stained with ethidium bromide which is visible under ultraviolet light. Protein is most commonly stained with Coomassie Blue or silver stains. Many bands are visible, with each corresponding to molecules of a certain size. Such stains bind to all molecules, and staining intensities are proportional to the mass of material in the bands. Detection of specific molecules requires additional techniques (see below).

Restriction digestion of DNA

Restriction enzymes
Restriction endonucleases are enzymes obtained from microorganisms that cleave DNA at specific sequences.

Restriction enzymes scan the length of a DNA molecule, until they encounter their specific recognition sequence, where they make one cut in each of the two sugar-phosphate backbones of the double helix.

The type II restriction enzymes employed for recombinant DNA technologies cut at short, usually palindromic recognition sequences to generate either blunt or sticky ends (Fig. 7.2). The recognition sites usually consist of four, six, or eight base pairs; sites for 8 base cutters are very infrequent, with those for 4 base cutters being fairly common.

Electrophoresis of restriction digested DNA

The digestion of human DNA with a six base cutter produces many fragments with an average size of about 4 kb. As the fragment sizes vary about this mean, digested human DNA appears as a 'smear' on a stained agarose gel (Fig. 7.3A).

Regular gels are limited in the size of fragment that they can separate, with a maximum of around 30 kb. Larger fragments can be resolved using 'pulsed field gel electrophoresis' (PFGE). Alternating the direction of the electric

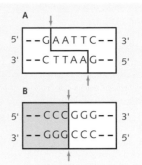

Fig. 7.2 Sticky and blunt ends after restriction enzyme digestion. (A) *Eco*RI produces staggered termini ('sticky ends'). (B) *Sma*I produces blunt ends. Adapted from Mueller and Young, 2001.

field allows fragments well over 1 Mb to be separated out. This specialized technique requires days, rather than the hours of standard gel electrophoresis.

Southern blotting

For detection of specific sequences (e.g. a gene) in genomic DNA, restriction digested DNA is denatured and transferred from the gel to a nylon membrane (Fig. 7.4). A probe comprising the sequence of interest is labelled with a radioisotope or chemical, denatured, hybridized to the target DNA on the blot, and then detected. The blot can be used in many consecutive hybridization experiments without the need to repeat the digestion and electrophoresis. While historically,

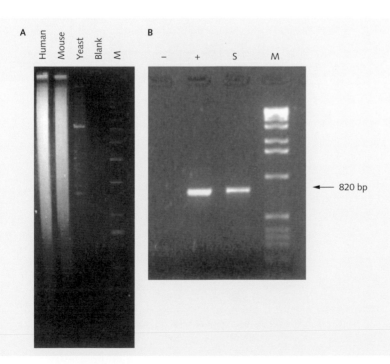

Fig. 7.3 Electrophoresis of genomic DNA. (A) Genomic DNA cut with 3 restriction enzymes that recognize different base pair sequences; the DNA appears as a smear of fragments. The marker lane (M) contains fragments of known size. (B) PCR products. An 820 bp band results because the primers are separated by this amount. The lane marked '–' is the negative control that contains all the components necessary for PCR except DNA. The lane marked '+' is the positive control, which is known to contain the sequence to which the primers anneal. The 820 bp band is seen in the sample lane (S), suggesting that this clone also includes the sequence to which the primers anneal. 'M' represents the marker lane of fragments of known sizes. Courtesy of Dr Steve Howe.

Fig. 7.4 Restriction map (A) and Southern blotting (B) showing diagnosis of sickle-cell disease: a point mutation in the β-globin gene that destroys a restriction site for *Mst*II. Genomic DNA is digested with *Mst*II and run on an electrophoresis gel. This separates the fragments, making the small ones travel further. The DNA is stained with ethidium bromide and visualized under ultraviolet light. The DNA is transferred onto a nylon membrane by Southern blotting. (1) Gel soaked in NaOH solution to denature DNA. (2) Filter paper under gel soaks up NaOH. (3) Dry filter paper sucks NaOH solution upward through gel and membrane, transferring DNA from gel to membrane. (4) DNA is fixed to the nylon membrane. Labelled probes hybridized to the Southern blot can be used to detect the β-globin gene. The mutant allele is detected as a single large *Mst*II fragment. Normal genes have two smaller *Mst*II fragments.

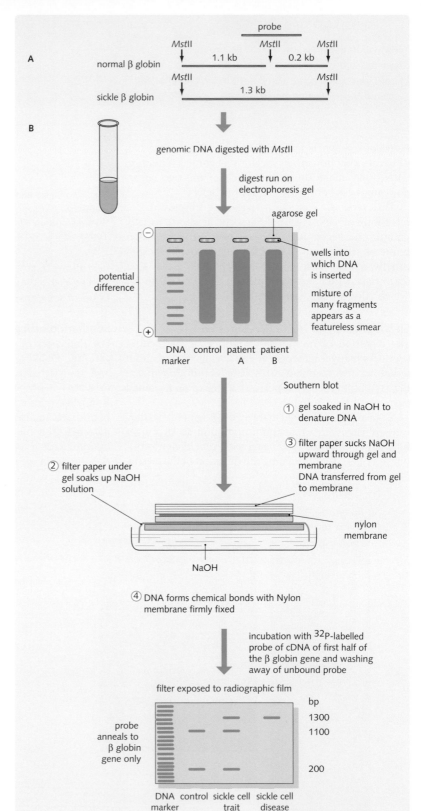

DNA sequence variants (RFLPs – see p. 120) were analysed in this way, they are now typed using PCR (see p. 110).

Northern blotting

Northern blotting is a technique used to study gene expression. It is analogous to Southern blotting, except that molecules of mRNA (rather than DNA) are separated by electrophoresis and transferred to a nylon membrane, which can then be used in hybridization experiments.

Molecular cloning

The identification of a set of overlapping clones spanning the genome was fundamental to the human genome project, and ultimately allowed the entire genome sequence to be pieced together. The availability of the genome sequence means that cloning is no longer central to human molecular genetics, but the methods are still used for some purposes.

The basics of cloning

In molecular cloning, DNA is introduced into a host (e.g. *E. coli*), and is replicated alongside the host's own DNA. The human DNA is combined with a 'vector' that contains specific sequences that are competent for replication in the host and that allow selection for host cells containing the vector.

In the simplest case, both insert and vector DNA are cut with the same restriction enzyme to produce molecules that have compatible sticky ends (Fig. 7.5). The vector is treated (de-phosphorylated) so that its free ends are unable to re-ligate together. The vector and insert sequences are mixed together and incubated with DNA ligase, which catalyses the formation of phosphodiester bonds between molecules of double-stranded DNA, joining the fragments together.

The ligation products are incubated with 'competent' host cells (some of which take up DNA) and are plated onto a selective media, on which only 'transformed' cells containing vector sequence can grow (Fig. 7.5). Many colonies grow, each of which is clonal, i.e. originates from a single cell (which was transformed by a single molecule of DNA). Special techniques (e.g. 'blue/white selection') may be used to identify recombinant clones (those that include insert DNA). In addition to the desired recombinant molecule, ligation may produce other events (e.g. recircularized vector). To identify the desired clones, several colonies are grown to produce DNA for analysis.

Growth of the host cells massively increases the amount of the insert DNA. If the inserted DNA came from a complex mixture (e.g. the human genome) cloning results in purification of the part of the genome

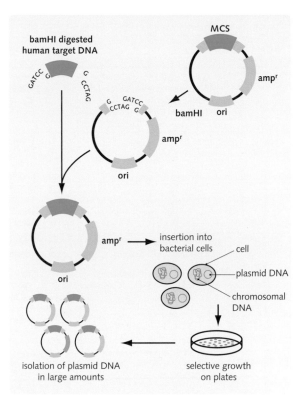

Fig. 7.5 Cloning. The plasmid contains an origin of replication (ori) and an antibiotic resistance gene (amp^r) that enable selection of transformants (bacteria that have taken up the plasmid); the multiple cloning site (MCS) contains several restriction sites to facilitate cloning of different fragments. Insert and vector DNA are prepared by restriction digestion to produce compatible ends (e.g. BamHI). Vector and insert are ligated and used to transform competent cells. Cells are plated onto media containing ampicillin, so that only transformed cells grow. Many different clones are produced, each containing a different fragment of DNA.

inserted into the vector. Thus molecular cloning effectively combines purification and amplification.

Host-vector systems

Various different systems have been developed for different purposes (Fig. 7.6). The most widely used host is *E. coli* for which several different types of vector have been developed to accommodate different sizes of inserted DNA. Yeast has also been used as extremely large inserts can be cloned into it.

DNA libraries

DNA libraries are populations of clones, each carrying an insert, which together, represent part or all of an organism's genome (genomic libraries) or complement of mRNA transcripts (cDNA libraries). Because RNA cannot be cloned directly, it is first reverse transcribed into

Fig. 7.6 Vector-host systems used in DNA cloning. Of these, plasmid and BAC are now the most widely used.

Vector	Origin	Features	Size of insert accommodated
Plasmid	Circular double stranded DNA in the cytoplasm of bacteria; undergoes replication with the bacterial genome and is passed on through generations; can be regarded as bacterial parasites	Origin of replication; antibiotic resistance genes; restriction enzyme site, which can break open plasmid and allow DNA to be inserted	Less than 10 kb
Phage	Viruses that infect bacteria	Phage particles can be assembled *in vitro*; DNA of interest is fragmented and ligated to phage *cos* sites; it is mixed with packaging extract, which contains all the proteins needed for phage assembly; phage heads are filled with DNA between two *cos* sites and a phage tail is attached; the assembled phage infects the host cell	16 kb
Cosmid	A genetically engineered hybrid of a plasmid and a phage	Contains plasmid origin of replication, selectable marker and phage *cos* site	45 kb
BAC	Bacterial artificial chromosomes are based on the F-factor plasmid	The vector includes the F-factor origin of replication, a chloramphenicol resistance gene and a marker that allows positive selection of recombinants on media containing sucrose. BACs are introduced into *E.coli* cells by electroporation	100–300 kb
YAC	'Yeast artificial chromosomes' are genetically engineered units that can be replicated in yeast cells	Contain the three DNA sequences essential for yeast chromosome function: telomeres (*TEL*), origin of replication (*ARS*), centromere (*CEN*)	100–1000 kb

DNA, termed cDNA (for copy or complementary DNA). Note that cDNA libraries represent the transcriptome of the cells/tissue/organ of origin at the time that mRNA was isolated. Only a small fraction of the genome is represented in cDNA libraries. Introns are present in genomic DNA, but not in cDNA clones.

PCR

Polymerase chain reaction (PCR) is a means of amplifying short segments of DNA (2–3 kb using standard methods), and the technique has revolutionized molecular genetics. Its many uses include:

- probe preparation
- to produce DNA fragments for cloning (see p. 109)
- genotyping polymorphic markers (see p. 120)
- diagnostic detection of mutations
- amplification of very small amounts of starting DNA, for example in forensics.

Principles of PCR

PCR hinges on the manipulation of conditions so that a DNA polymerase enzyme repeatedly replicates a specific sequence of chosen DNA (Fig. 7.7). The amplification is exponential because in each cycle of PCR, the products of the previous reactions are used as templates. Each reaction includes a pair of primers that together flank the target sequences (Fig. 7.7). These are synthetic oligonucleotides, usually of 18–22 bases in length, and define the 'region' of DNA to be replicated. This is very useful as it is often desirable to amplify only a small section of a particular sample.

PCR in practice

There are five reagents essential for PCR.

Template DNA
The template DNA is the sample, which contains the target region that is to be amplified.

HINTS AND TIPS

Very little DNA is required for a PCR reaction. Sufficient DNA is routinely extracted from buccal cells, which can be painlessly scraped from the inside of the cheek. For diagnosis of genetic disease, even a single cell of a preimplantation embryo can provide sufficient DNA.

Primers
Primer design is key, as the sequences give PCR its specificity for amplification of target regions. Amongst various important considerations, the primers should be designed to have similar melting temperatures, and to be unable to anneal to each other.

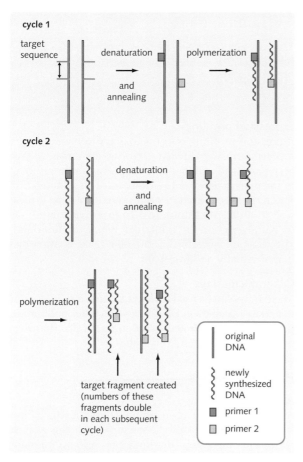

cycle 1

target sequence

denaturation and annealing

polymerization

cycle 2

denaturation and annealing

polymerization

target fragment created (numbers of these fragments double in each subsequent cycle)

	original DNA
	newly synthesized DNA
■	primer 1
□	primer 2

Fig. 7.7 The polymerase chain reaction (PCR). Each primer is complementary to the sequence on one strand, flanking the target. *Taq* polymerase catalyses the addition of nucleotides onto the 3' ends of both primers. The products of one cycle become the templates for the next.

Thermostable polymerases

Taq polymerase is a heat-stable DNA polymerase enzyme originally isolated from *Thermus aquaticus*, a thermophilic bacterium that inhabits hot springs. Like all DNA polymerases (see also p. 5) *Taq* polymerase:

- requires a primer to initiate synthesis
- synthesizes DNA in a 5'–3' direction.

Taq polymerase lacks 3'–5' (proofreading) exonuclease activity, having a misincorporation rate of about 1 in 10 kb (much higher after many cycles of PCR as errors are compounded!). If sequence fidelity is important, a proofreading polymerase with 3'–5' exonuclease activity, such as Phu polymerase (an engineered polymerase) can be used.

dNTPs

Deoxynucleotide triphosphates (dNTPs) are the substrates for *Taq* polymerase, from which the new strands of DNA are synthesized. The reaction should include adequate and equal amounts of dATP, dTTP, dCTP and dGTP.

Buffer

The buffer maintains the optimum pH and chemical environment for the polymerase enzyme and for specificity of primer annealing.

The reaction

Tubes containing the various reagents are placed in a 'thermocycler'. This rapidly heats and cools the reaction tubes in a cyclical manner, each cycle consisting of heating and cooling to three distinct temperatures.

Each cycle theoretically doubles the amount of DNA, although this is never achieved and for various reasons the reaction becomes less efficient as products accumulate. Typically 35 cycles are performed, after which about 10^{12} copies of the target sequence are present. This is sufficient DNA to be visualized on an agarose gel. The steps involved in each PCR cycle are summarized in Figure 7.8.

HINTS AND TIPS

PCR has become an essential tool for the diagnosis of infection and in infection control. It can provide rapid identification of organisms and is especially useful when the target organism is present in low numbers, or is difficult to culture *in vitro* (e.g. *C. pneumoniae* and *M. tuberculosis*); viral infections can be detected by PCR before seroconversion.

Fig. 7.8 A summary of a typical PCR reaction, with an approximate time scale.

Polymerase chain reaction		
Step 1	95 °C for 4 minutes	Initial denaturation step
Step 2	94 °C for 1 minute	Short denaturation period
Step 3	55 °C for 1 minute	Primer-annealing step
Step 4	72 °C for 1 minute	Elongation step
Step 5	Go back to Step 2, 30 times	Cycling steps
Step 6	72 °C for 10 minutes	Final elongation step
Step 7	Hold at 4 °C	

Controls

The exquisite sensitivity of PCR renders it vulnerable to contamination artefacts; tiny amounts of contaminating DNA can generate spurious results. For this reason every experiment should include a negative control containing all the reagents except for the target DNA. Amplification in this reaction shows contamination. It is good practice also to perform a positive control reaction including a standard DNA.

RT-PCR

This modification of PCR is used to study gene expression. Because *Taq* polymerase requires a DNA template, RNA is copied into DNA using the viral enzyme reverse transcriptase (RT). In the first step, RNA is incubated with this enzyme in the presence of dNTPs and an appropriate primer for synthesis of a strand of cDNA. The products are then used in a PCR reaction.

qPCR

Quantitative PCR (qPCR) is a technique used to simultaneously amplify and quantify target molecules. The most common use of this is quantitative RT-PCR for measurement of gene expression (qRT-PCR). The technique involves detection of the amplified DNA in real-time as the PCR reaction progresses, using either fluorescent dyes or labelled oligonucleotide probes.

HINTS AND TIPS

Confusingly, qPCR used to be known as real time PCR, also abbreviated as RT-PCR!

Microarrays

DNA microarrays (or 'gene chips') consist of hundreds of thousands of different specific oligonucleotide probes arrayed in a series of spots (or 'features'), fixed on a solid support. Target DNA is labelled with fluorescent dye, denatured and applied to the array under controlled conditions (Fig. 7.9). The amount of fluorescence at each feature gives a measure of the abundance of the target. DNA Microarray analysis allows the target to undergo many thousands of tests simultaneously.

SNP typing

Single nucleotide polymorphisms (SNPs) are genetic variants that occur when a single nucleotide in the genome is altered (see p. 120). SNP microarrays are capable of testing hundreds of thousands of SNPs in the human genome and have many uses including forensic testing, detecting predisposition to disease and genetic mapping (see p. 119).

HINTS AND TIPS

A single point mutation that occurs in more than 1% of the population is called a single nucleotide polymorphism.

Expression profiling

Expression profiling uses microarrays to measure gene expression, using labelled cDNA as the target. Microarrays are available that allow expression of every gene in the human genome to be measured simultaneously. This technique is used to study patterns of gene expression in different samples (e.g. to investigate responses to pathogens, or for molecular typing of tumours). The relative expression of different genes is often displayed visually on 'heat maps'.

Other microarrays

Microarrays can be used to detect a number of biological materials other than DNA, these include: antibodies, proteins, carbohydrates (glycoarrays), cells (transfection microarrays) and even tissues.

Protein analysis

Antibodies

Antibodies are circulating immune proteins which bind strongly and specifically to the target proteins against which they are raised. They underpin various important protein detection methods including:

- western blotting
- enzyme-linked immunosorbent assay (ELISA)
- immunohistochemistry.

Protein detection is usually by a two-layer method. The primary antibody probe binds specifically to the target protein. This is then detected using a labelled secondary antibody raised against all antibodies of the species from which the primary antibody was obtained (e.g. mouse, rabbit, goat); the label is often a fluorescent tag (e.g. Alexa dyes) or an enzyme (e.g. horseradish peroxidase). Enzyme labels are detected using substrates that produce a coloured or fluorescent product, or light. Such methods enhance the sensitivity of antibody-based detection.

Gel electrophoresis and western blotting

Western blotting is used to detect the presence of a specific protein in a sample and is the protein equivalent of northern and Southern blotting (see p. 107).

The sample proteins are first denatured and reduced and electrophoresed on a sodium dodecyl sulfate (SDS)-containing polyacrylamide gel, separating them by molecular weight. The proteins are transferred to a nitrocellulose or PVDF membrane and probed using

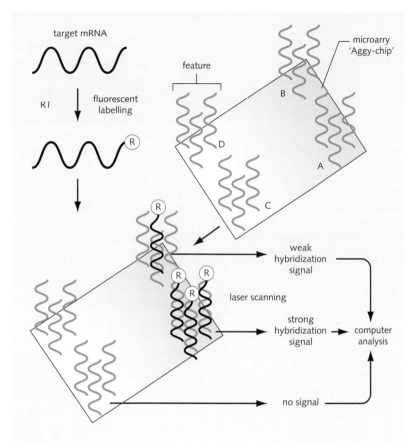

Fig. 7.9   The principles of expression microarray analysis. The microarray is prepared by fixing or synthesizing groups of probes (features) in known positions on a solid support (chip). mRNA sample is reverse transcribed to generate fluorescent cDNA 'target', which is hybridized to the array. The microarray is washed to remove unbound molecules (not shown). Many molecules of cDNA are complimentary to feature A, producing a strong signal on laser scanning. A little cDNA is complimentary to feature B, producing a weak signal. None of the sample molecules are complimentary to features C or D.

a specific antibody to the protein under investigation, as described above. Other gel systems allow separation by size and shape (native PAGE) or charge (isoelectric focussing).

HINTS AND TIPS

Southern blotting was named after its developer Ed Southern. Northern blotting and western blotting were named as a pun. A handy mnemonic for remembering which molecules each hybridization technique detects is **SNOW DROP**:

Southern **D**NA
Northern **R**NA
O   **O**
Western **P**rotein

The Os are zeros, since there is no eastern blot.

ELISA

Enzyme-linked immunosorbent assay (ELISA) is used to detect the presence of an antigen or antibody in a sample. Typically the protein of interest is captured on an antibody-coated surface (usually a polystyrene microtiter plate), and detected using a second antibody specific for the protein of interest. ELISA is quicker, consumes less of the sample and is more quantitative than western blotting, but gives no information about molecular weight.

Cytogenetics

Cytogenetics is the study of chromosomes and their abnormalities. At metaphase of the cell cycle, the chromosomes are condensed and can be spread on a slide, stained and viewed by microscopy. G-banding and FISH enable individual chromosomes to be identified, FISH can also identify specific sequences within them, while CGH allows for loss or amplification of specific chromosome regions to be detected, as can MLPA.

Preparation of chromosome spreads is achieved using chemicals (e.g. colchicine or colcemid) that arrest the cell cycle in metaphase.

G-banding

G-banding is the most commonly used chromosome staining technique and is the mainstay of cytogenetic diagnosis. Chromosomes are subjected to a controlled protein digestion with trypsin and stained with Giemsa. This results in a pattern of dark and light bands that is specific for each chromosome, allowing them to be identified (Fig. 7.10). It is used in the diagnosis of:

- monosomies and trisomies
- translocations
- large deletions and insertions.

For many of its applications, G-banding is being superseded by FISH.

FISH

Fluorescence *in situ* hybridization (FISH) is a method of visualizing specific regions of metaphase or interphase chromosomes. Spread chromosomes are fixed to a microscope slide and the DNA is denatured. It is then hybridized with fluorescently labelled probe DNA that binds specifically to a complementary sequence. The region of the chromosome where hybridization has occurred can be visualized. FISH probes may bind to a single genomic sequence (Fig. 7.11) or stain an entire chromosome ('chromosome paints'). Simultaneous use of different fluorescent labels allows multiplex fluorescence *in situ* hybridization (M-FISH) to detect multiple targets, and 'spectral karyotyping' (SKY) in which each chromosome appears in a different colour. In addition to being used in research, FISH is used diagnostically to identify a variety of abnormalities including:

- monosomies and trisomies
- translocations
- microdeletions and insertions.

FISH methodology is more sensitive than G-banding.

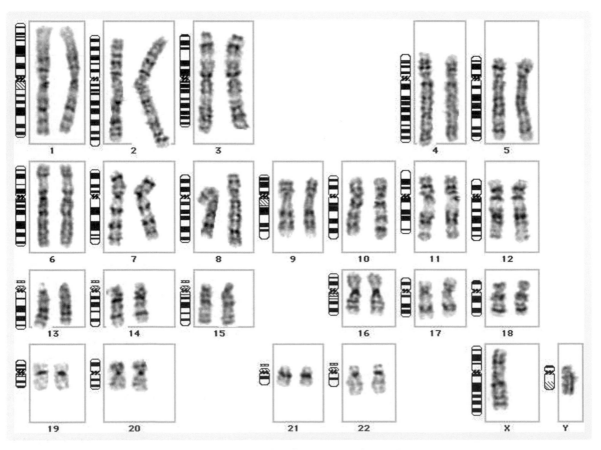

Fig. 7.10 G-banding: a karyotype for normal human male cells. (Courtesy of Dr Linda E. Ritter.)

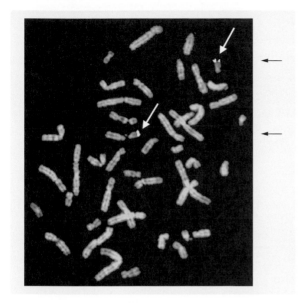

Fig. 7.11 Fluorescence *in situ* hybridization (FISH). Using a single copy probe. Although the probe is said to be single copy, this refers to the haploid genome. On the metaphase spread, two pairs of dots are visible (arrowed) that correspond to hybridization of the biotin-labelled probe to both chromatids on a pair of homologous chromosomes. (Courtesy of Dr Paul Scriven, GSTT.)

CGH

Comparative genomic hybridization (CGH) is a two-colour FISH technique that allows chromosomal losses and duplications to be quantified. Equal amounts of differentially fluorescently labelled test genomic DNA, frequently tumour DNA, and normal reference DNA are mixed together and hybridized to normal metaphase spreads. Areas of chromosomal duplication in the test sample will hybridize, with excess quantities of its label, to the corresponding region of the metaphase spread. Areas of deletion in the test sample will lead to its under-hybridization with the metaphase spread, which will hybridize the labelled normal DNA instead (Fig. 7.12). CGH is used clinically to:

- screen chromosomal copy number changes in tumour genomes
- study tumours that do not yield sufficient metaphases for other forms of analysis
- study archival material allowing correlation of chromosomal aberrations with the clinical course.

A more powerful modification of the CGH technique, known as array CGH, exploits new microarray technology using several thousand probes, rather than metaphase chromosomes. This can detect much smaller areas of deletion and amplification (5–10 kb).

MLPA

Multiplex ligation-dependent probe amplification (MLPA) is a method that is replacing FISH for the detection of many disorders. It detects copy number variation in genomic sequences at high resolution (e.g. single exon deletions) and, unlike FISH, is easily amenable to multiplexing. For each locus of interest, two oligonucleotide probes are synthesized that anneal to the target sequence end to end. Annealed probes are joined by a ligase and then PCR is used to amplify the probe ligation product. Variation in the length of the original probes allows testing of up to 40 loci in one reaction (Fig. 7.13). The relative amount of product is proportional to the copy number so can be used to detect deletions or duplications. Allele specific probes can be used to detect single base changes as ligation is prevented by the mismatch between the genomic and probe sequences (Fig. 7.13).

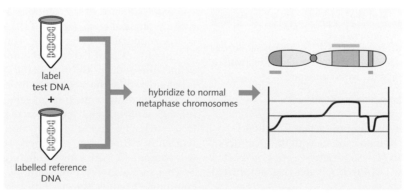

Fig. 7.12 Comparative genomic hybridization (CGH). Fluorescently labelled test sample DNA (FITC stained; green–grey in this diagram) and reference DNA (rhodamine stained; red) are hybridized to normal metaphase chromosomes. Equal co-hybridization of the green and red probe indicates a stable copy number at that point on the chromosome. DNA losses in the test sample see only reference DNA binding, leading to a predominance of the red fluorophore, while DNA gains lead to excess test sample DNA binding, out-competing the reference DNA, and a predominance of the green fluorophore.

Fig. 7.13 Multiplex ligation-dependent probe amplification (MLPA). MLPA probes consist of two oligonucleotides: a short synthetic oligonucleotide, and a longer oligonucleotide from phage M13. Each long oligonucleotide has a different stuffer sequence. Following hybridization, ligation is achieved using ligase. Only perfectly matched probes will ligate. As the amplification product of each probe has a different size, varying from around 130–480 bp, the relative amounts of each product reflects the relative copy number of that target sequence.

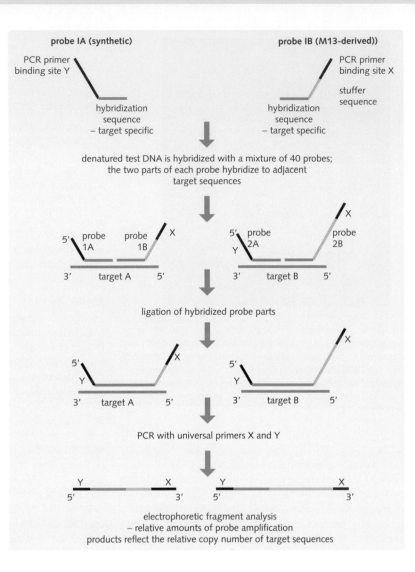

THE HUMAN GENOME PROJECT

The human genome project (HGP) is an international cooperative research effort to investigate the human genome in its entirety. Its main aims are:

1. Mapping human genes and markers
2. Sequencing the genome
3. Functional analysis and post-genomic genetics
4. Comparing the human genome with the genomes of model organisms (*E. coli, S. cerevisiae, C. elegans*, etc.)
5. Developing new DNA technologies (e.g. automated sequencing)
6. Developing bioinformatics (systems for collecting, storing, and disseminating information generated by the project)
7. To explore the ethical, legal and socioeconomic context of the findings.

The base sequence of the genome is the most detailed type of physical map (see p. 119). The ultimate aim is the production of a single, continuous sequence of bases for each of the human chromosomes, and the delineation of the position of all the genes. The sequence is determined by the steps described in Figure 7.14.

A 'rough draft' of the human genome was completed in 2003, and final HGP papers were published in 2006. It is envisaged that the correction of minor errors (currently estimated at 1 in every 10000 nucleotides) will continue for the foreseeable future.

There are also a number of arms of the project still continuing to study aspects of the genome. For example, 'HapMap' aims to construct high-density SNP maps, which will allow researchers and clinicians to:

- search for and isolate specific disease causing genes – single gene disorders

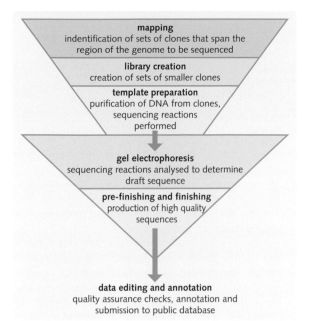

Fig. 7.14 Stages involved in the human genome project. The processes involved in mapping the human genome can be thought of like two large sieves. The first produces small, accurate clones; the second is concerned with increasing sequencing specificity.

- compare SNP patterns from populations with a given multifactorial disease (i.e. cancer, heart disease, type two diabetes) with those from unaffected individuals
- guide the development of new drugs targets.

Overlapping clones

The human genome is a vast entity of approximately 3 billion base-pairs. In order to be studied it had to be broken down into a number of manageably-sized over-lapping clones to be sequenced. Such an array of overlapping DNA segments is termed a 'contig'. The genome sequence was assembled by piecing together the sequences of overlapping clones.

Sequencing

Sequencing is the determination of the order of bases in a molecule of DNA. There are a variety of methods that may be used for sequencing:

- chemical cleavage (Maxam and Gilbert method)
- chain termination (Sanger sequencing)
- pyrosequencing.

Automated Sanger sequencing was used almost exclusively for the initial sequencing of the human genome.

Sanger sequencing method

The principle of Sanger sequencing hinges on DNA polymerase being unable to extend a growing DNA strand once a nucleotide analogue (a ddNTP) that lacks the 3' hydroxyl group has been incorporated (Fig. 7.15).

Historically, Sanger sequencing was achieved using radiolabelled dNTPs. Using this method, for each piece of DNA sequenced, four parallel reactions are set up,

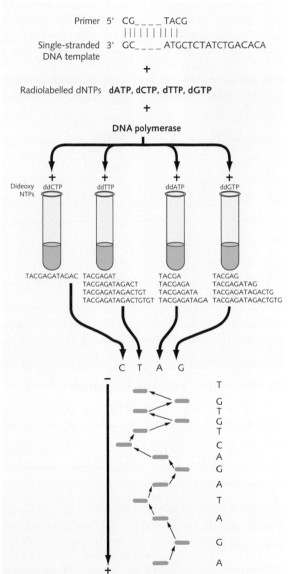

Fig. 7.15 The Sanger method of determining the nucleotide sequence of a segment of cloned DNA. Dideoxy NTPs are chain terminating bases; four reactions are set up, each with one ddNTP. The mixture of dideoxy-terminated molecules are electrophoresed on a polyacrylamide gel to determine the sequence. (Adapted from Mueller and Young, 2001.)

each with one of the four dideoxynucleotide triphosphates (ddNTPs):

- template DNA (DNA to be sequenced)
- the sequencing primer
- dNTPs (dATP, dTTP, dGTP and dCTP, one of which is radiolabelled)
- ddNTP (one of ddATP, ddTTP, ddGTP or ddCTP)
- DNA polymerase.

Extension of the DNA chain continues until a ddNTP (a 'chain terminator'), is incorporated. Since each reaction contains all four dNTPs (onto which bases can be added), a population of newly synthesized DNA strands result that vary in length according to how quickly a ddNTP was encountered at random. These can be separated by electrophoresis on a polyacrylamide gel, and the radioactive bands can be visualized by exposure to X-ray film (Fig. 7.15).

HINTS AND TIPS

Although rarely used, the methodology behind four-pot Sanger sequencing is a popular pre-clinical exam question!

The HGP was mostly undertaken using an automated technique which utilized ddNTPs, each labelled with a different fluorochrome (known as dye-deoxys). For each DNA sequence, only one reaction is required, with the electrophoresis being performed in capillaries with 4 colour optical detection. This technique is now being overtaken by even faster 'next generation' technologies, which avoid the need for gel electrophoresis.

The reference genome sequence

The reference genome is a digital copy of the base sequence of the human genome. During the HGP it was contributed to by a large number of international teams and is publically available online. It is continually being updated as small gaps are filled, and errors corrected.

The reference genome was assembled from a number of donors, and so does not perfectly represent the genome of any one person. However, it now serves as a framework on which new genomes are compiled, enabling them to be assembled much quicker than the initial HGP. A number of individuals' genomes have been sequenced, including that of James Watson, one of the 'fathers of DNA'.

In silico techniques: bioinformatics

Bioinformatics is an integration of mathematical, statistical and computer methods to analyse molecular data. With the advent of high throughput, automated

techniques, the storage of raw data in databanks allows for comprehensive study of:

- normal biological processes
- abnormal biological processes and disease
- modelling biological systems
- tailored drug discovery.

Methods include:

- DNA informatics, including sequence analysis, open reading frame (ORF) analysis and gene scanning
- genomics, the application of DNA informatics to analyse the genome
- protein informatics, using protein sequences to model structure and function
- proteomics, the application of protein informatics to analyse the protein complement of the genome
- metabolomics, defined as *'the quantitative measurement of all low molecular weight metabolites in an organism's cells at a specified time under specific environmental conditions'*, looks not only at all the end products of gene expression, but does so in a particular physiological background or developmental state.

Next generation sequencing

A huge amount of technology and understanding were developed as a result of the HGP. The demand for low-cost, high throughput sequencing has lead to the development of next generation technologies such as ion semiconductor, nanoball and massively parallel signature sequencing. A number of new projects have emerged.

- 1000 genomes project – aims to sequence the genomes of over 1000 people from across the world, in order to establish a detailed catalogue of genetic variation.
- The 'thousand-dollar genome' – the possibility of reducing costs to a point where sequencing an individual's entire genome becomes routine, an enormous step towards the realization of 'personalized medicine'.

Transcriptomics

The transcriptome is the full complement of RNA molecules in a population of cells, including mRNA, rRNA, tRNA and other non-coding RNA.

A surprise revealed by the HGP is that humans appear to require only 20 000–25 000 protein-coding genes, which is only double the number that flies and worms have. However, human genes are more complex, and processes such as alternative splicing greatly increase the number of proteins produced by widening the transcriptome (Fig. 7.16).

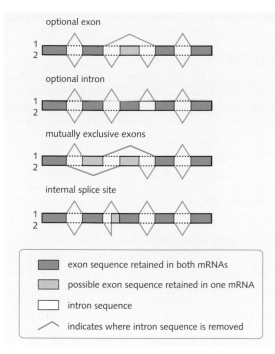

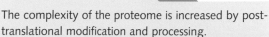

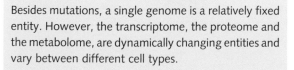

Fig. 7.16 Alternative splicing. Four patterns of alternative splicing are shown; in each, a single primary RNA transcript is spliced in two alternative ways to produce two distinct mRNAs (1 and 2), though greater complexity patterns exist.

HINTS AND TIPS

Besides mutations, a single genome is a relatively fixed entity. However, the transcriptome, the proteome and the metabolome, are dynamically changing entities and vary between different cell types.

The human proteome and proteomics

The proteome is the complete set of proteins by an organism, though in practice, proteomics usually focuses on the proteins expressed in single cell type or organ. Proteins are characterized and sequenced in a number of ways.

2D gel electrophoresis

2D electrophoresis separates proteins by two properties. Proteins are first separated by isoelectric point and then, in a perpendicular direction, by mass; the proteins appear as an array of spots on the gel. Because few proteins are similar in both these properties, molecules are well separated. Comparison of 2D electrophoresis patterns allows identification, e.g. of proteins enriched in disease. Such spots can then be cut out of gels and identified by mass spectroscopy.

Mass spectroscopy

Mass spectroscopy can be used to characterize and sequence proteins. Whole proteins are vaporized, ionized and separated according to their mass-to-charge ratio. They are then quantified by a detector and analysed to provide information about the protein's composition in a process known as 'top-down' analysis. Peptide mass fingerprinting is where proteins are partly enzymatically digested before spectroscopy, and are recognized by the characteristic pattern of peptides produced.

HINTS AND TIPS

The complexity of the proteome is increased by post-translational modification and processing.

GENETIC MAPS AND IDENTIFICATION OF DISEASE GENES

It is vital to understand which genetic abnormalities cause disease for:

- definitive diagnosis
- more accurate assessments of risk or prognosis
- presymptomatic diagnosis
- prenatal diagnosis.

Furthermore, understanding of the molecular basis of disease identifies new drug targets, enables rational drug design and facilitates the production of animal models of disease on which therapies can be tested.

The first step in identification of pathogenic mutations is mapping.

Genetic mapping

Genomes can be mapped in two ways. Maps constructed using information about transmission through generations of genes or other characters are called 'linkage maps'; 'physical maps' are based on physical properties of chromosomes and DNA. The linkage and physical maps have been aligned to each other.

Linkage maps

Linkage mapping of the human genome is based on the probability of recombination between paternally and maternally derived chromosomes at meiosis (see p. 98):

- loci are assigned to linkage groups (genetic map equivalent of the chromosome)
- loci are mapped within linkage groups with respect to each other

- map distances are quoted in recombination units (centimorgans).

One centimorgan (1 cM) is equivalent to a 1% chance of recombination. Recombination is more frequent in female meiosis, so the male and female maps are different, with markers appearing further apart on the female version. Some areas of the genome are more prone to recombination than others, so there is not a perfect alignment between genetic and physical maps.

Linkage mapping is fundamental to positional cloning (see p. 120). The most recent high-resolution genetic maps have polymorphic markers spaced at intervals of less than 1 cM.

HINTS AND TIPS

Genetic and physical map distances are not perfectly correlated, but as a rule of thumb, 1 cM is equivalent to 1 Mb.

Physical maps

Physical maps locate genes at physically defined locations, and give distances in bp, kb or Mb. Physical maps were constructed using restriction mapping, sequencing and cytogenetics to analyse overlapping clones, anchored to specific named chromosomes. The ultimate physical map is the full human genome sequence.

HINTS AND TIPS

The data generated by the human genome project have greatly simplified positional cloning. When the approach was first suggested in the 1980s, individual research groups had to identify polymorphic markers themselves with which to establish linkage. Once linkage was found, they had to construct their own physical maps and contigs of overlapping clones before they could even begin to look for the disease-causing gene. Now, all this is freely available thanks to the efforts of the human genome project.

Positional cloning

Positional cloning (reverse genetics) is a method of identifying a disease gene when nothing is known about the nature of the corresponding protein. This approach was used to identify the genes responsible for cystic fibrosis and Huntington disease.

In positional cloning, the gene responsible for a condition is identified from knowledge of its position in the genome from linkage analysis of polymorphic markers in families in which the disease is segregating (see 'genetic linkage analysis', p. 121) and/or physical map data, e.g. chromosomal anomalies segregating with the disease.

Polymorphic markers

A marker is any Mendelian characteristic used to follow transmission through a pedigree. DNA polymorphisms that have a known position on the genetic map of the human genome are used as markers in linkage studies.

- DNA polymorphisms are inherited differences in DNA between healthy people.
- They frequently arise in non-coding regions.
- Examples include microsatellites, single nucleotide polymorphisms (SNPs) and restriction fragment length polymorphisms (RFLPs).

Microsatellite markers

Variable number tandem repeats (VNTRs) are short nucleotide sequences up to 100 bp long organized into clusters of 'tandem repeats'. A useful family of VNTRs are microsatellites which are characteristically between 2 and 6 bp in length. Microsatellite repeats rarely occur within coding sequences and are well suited as markers for genetic linkage analysis.

- They can be detected and typed by PCR.
- They are extremely polymorphic.
- Tens of thousands of microsatellite polymorphic loci exist throughout the genome.

An example of the results obtained after PCR with a single pair of primers that spans a microsatellite is shown in Figure 7.17. Multiplex microsatellite typing allows several loci to be typed in a single analysis.

HINTS AND TIPS

Each allele of a microsatellite may give rise to several bands on a polyacrylamide gel, which makes interpreting a single lane difficult. To identify which bands are parts of the same allele, you must compare the bands across all the lanes (see Fig. 7.17).

SNPs

Single nucleotide polymorphisms (SNPs) and restriction fragment length polymorphisms (RFLPs) are DNA sequence variations that occur when a single nucleotide in the genome sequence is changed. They occur approximately once in every 200–500 nucleotides, most of which lie in non-coding regions of the genome, and are thought to be silent.

They are a useful type of marker as they occur frequently, and millions are known. Though less polymorphic than microsatellites, they can be easily detected using hybridization, and it is possible to detect many thousands simultaneously using microarrays (see

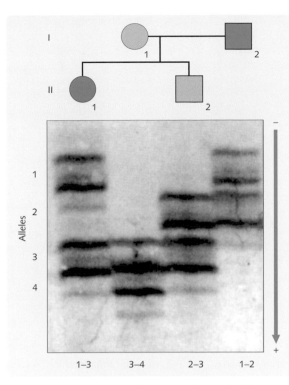

Fig. 7.17 An autoradiograph showing the results of a PCR using primers that span a microsatellite repeat. The microsatellite is closely linked to a gene that causes a dominant disorder and allele 1 is segregating with the condition. Note that each allele gives rise to multiple bands, which is due to an artefact of PCR. (Reproduced from Mueller and Young, 2001.)

p. 112). For large-scale analyses, SNPs are preferable to microsatellite markers.

> **HINTS AND TIPS**
>
> Microsatellites and SNPs are scattered throughout the genome.
>   Genetic fingerprinting used for paternity testing and in forensics, is usually conducted by PCR analysis of a panel of microsatellites.

Genetic linkage analysis

Linkage
Linkage analysis uses pedigree data to determine whether loci are linked and to estimate the recombination fraction (see below). At meiosis, homologous chromosomes exchange segments before separating into two daughter cells.

  Linked markers segregate together in meiosis more frequently than expected by chance because they lie close together on the same chromosome (Fig. 7.18). Crossovers are statistically unlikely to form between

markers that are close together. However, if markers are sufficiently far apart it is likely that a crossover will form between them and recombination will occur.

>
> **HINTS AND TIPS**
>
> For mapping of disease genes, the aim is to establish linkage of the disease phenotypes with markers of known position.

Recombination
Recombination occurs as a result of the formation of chiasmata (crossovers) during meiosis. Recombinants are children who inherit a different combination of alleles at two loci compared with the combination found in the gametes that made the parent.

  The recombination fraction (RF) is the proportion of recombinants. For unlinked loci it is 0.5; for linked loci it lies between 0 and 0.5. Recombination fractions are used to produce genetic maps (Fig. 7.19). (1 cM is defined as a 1% chance of recombination, which is equivalent to an RF of 0.01.)

The log of odds score
The random assortment of chromosomes at meiosis (see p. 98) means that unlinked alleles on different chromosomes will segregate together on average 50% of the time (i.e. RF = 0.5). Thus, even in large families with many meioses, markers may appear to be linked because they have repeatedly segregated together by chance. A log of odds (LOD) score is a statistical estimate of whether two loci are likely to lie near each other on a chromosome, and are, therefore, likely to be inherited together. A LOD score of three or more implies that the two loci are linked.

Family selection and pedigree analysis
The first step in identifying a disease gene is finding sufficient families with the gene to identify linkage. The role of the clinician is vitally important, as it is imperative that all affected members of the family are identified and that all the families included have the same disease. The mode of inheritance (recessive, dominant, etc.) should also be considered, as this information is required for linkage analysis. Mutations in known disease genes should be excluded.

> **HINTS AND TIPS**
>
> Genetic diseases can result from mutations in different genes in different families. This is termed 'genetic heterogenicity'.

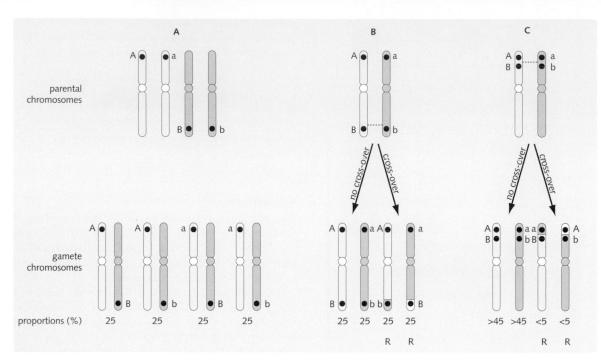

Fig. 7.18 Segregation at meiosis of alleles at two loci. Recombinants are marked (R). (A) The loci are on different chromosomes. (B) The loci are on the same chromosome, but widely separated (>50 cM apart). (C) The loci are closely adjacent and so mostly segregate together (i.e. linked). (Adapted from Mueller and Young, 2001.)

Establishing linkage to a chromosome

To establish linkage to a specific chromosome, all members of the family are genotyped with respect to about 400 polymorphic markers.

- These markers are nearly always microsatellites.
- They are freely available through the HGP.
- They are scattered throughout the genome.
- They have known positions on genetic and physical maps.

The data generated are reviewed to determine whether any of these markers is segregating with the disease (Fig. 7.20).

Fine genetic mapping

Once linkage to a particular chromosome or chromosomal region has been identified, all the members of the family are genotyped with respect to a further set of more densely packed markers that map to the appropriate chromosomal region. This aims to:

- confirm linkage
- narrow the region of the genetic map in which the disease gene can be.

Even before disease-causing mutations are identified, linked genetic markers (5 cM or less) can be used to track the inheritance of a disease gene for diagnostic purposes. For positional cloning, the region of interest must be flanked by markers no more than ~1cM apart.

From linked genetic markers to disease gene

The output of the HGP means that, once the relevant genetic region has been identified, researchers and diagnosticians can go straight to the HGP sequence maps, though many genes will normally reside in the candidate region. These are then investigated as candidate disease genes for identification of pathogenic mutations.

Mutation analysis

If mutation is responsible for a genetic disease, then the genotypes and phenotypes should be concordant, i.e. (for a recessive disorder) homozygous mutant genotype in affected individuals; normal or heterozygous genotype in unaffected family members and the population at large.

Rapid methods for screening of mutations include:

- single strand conformation polymorphism (SSCP) analysis
- heteroduplex analysis
- direct sequencing (the current method of choice).

Validation requires further evidence, e.g. multiple different mutations in the same gene in different families. There should be biological evidence connecting the gene with the specific disease, e.g. involvement in a disrupted biochemical pathway, gene expression in tissues affected by the disorder, or an animal model.

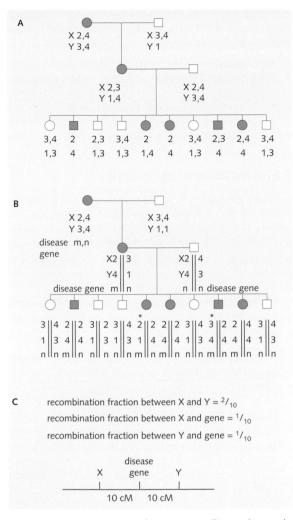

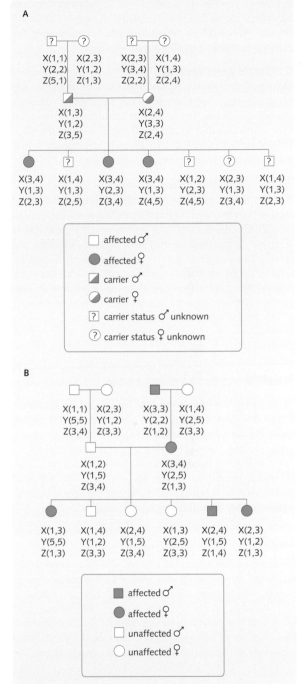

Fig. 7.19 Genetic mapping of two microsatellite markers and a dominant disease gene. (A) Family members are genotyped with respect to two microsatellite markers (X and Y). (B) The markers are rearranged according to parental inheritance. (C) The recombination fraction for each pair of markers is calculated separately using the children's chromosomes and the genetic map is constructed (1% recombination fraction is equal to 1 cM). *, recombinant allele; m, mutant disease gene; n, normal disease gene; X and Y, microsatellite markers; 1–4, allele sizes. (Courtesy of Dr Kathy Mann.)

Other methods of identification of disease genes

Various other strategies have been employed to identify the genes that are mutated in genetic disease, and new approaches are emerging. Some important strategies are discussed below.

Functional cloning

Functional cloning is the identification of a gene responsible for a disease based on knowledge of the underlying molecular defect. If the nature of the

Fig. 7.20 Identifying linkage between polymorphic markers and disease genes. Three polymorphic markers are used (X, Y and Z) that map to three different chromosomes (2, 3 and 5, respectively). (A) In this pedigree showing autosomal recessive inheritance, the disease is segregating with the '3' and '4' alleles of marker X, suggesting that the disease maps to chromosome 2. (B) In this pedigree showing autosomal dominant inheritance, the disease is segregating with the '1' allele of marker Z, suggesting that the disease maps to chromosome 5.

deficient protein is known, it is possible to use biochemical methods to identify the gene. This approach was used to define the genes responsible for phenylketonuria and sickle-cell anaemia.

Candidate gene approach

The candidate gene approach is the use of previously isolated genes as disease gene candidates through knowledge of the physiology of the diseased tissue, or the nature of other genes involved in similar diseases. This approach was used to identify rhodopsin mutations associated with retinitis pigmentosa.

Exome and genome sequencing

With high throughput sequencing and knowledge of the range of normal and pathogenic variations, it is possible to identify disease-causing mutations in individuals directly. These approaches can readily identify the genetic basis of disease in patients in which the mutation(s) occur in known disease genes, and have the potential for identification of candidate novel disease genes. Most pathogenic mutations are within or near exons. Sequencing of all exons (the 'exome') provides a cheaper and more focused approach than whole genome sequencing.

Multifactorial diseases

Many common disorders have a genetic component. While linkage mapping has successfully identified the genetic basis of many monogenic diseases, it does not have adequate power to identify multiple gene effects responsible for multifactorial diseases (see p. 137). However, recent advances in high-throughput SNP genotyping technology have allowed large-scale population (rather than family) based studies.

GWAS

Genome-wide association studies (GWAS) examine the genomes of many individuals, and compare their genetic differences to phenotypic traits. In humans, SNP genotyping has examined over 200 diseases and traits, and has proven to be powerful in the identification of a number of groups of genes associated with diseases including diabetes and inflammatory bowel disease.

GENETICALLY ENGINEERED THERAPEUTICS

Advances in molecular genetics have lead to the development of a number of treatment modalities. Advancing knowledge has identified new targets for conventional small molecule therapies, and individual genotyping is beginning to allow therapeutics to be highly tailored to individuals at a molecular level (personalized therapy). Further, a number of new treatments modalities have emerged as a result of genetic engineering.

- Gene therapy (see p. 150).
- Genetic modification of cells and transplant technologies.
- Recombinant therapeutic proteins.
- Genetically engineered vaccines.
- Genetically engineered antibodies.

Recombinant proteins

In many diseases that result from protein, enzyme or hormone deficiencies, the deficient product can be externally produced and replaced. Historically, these products were isolated from animals, healthy people or cadavers, and purified. However, this posed a number of problems including expense, immuno-incompatibility and disease transmission.

The healthy human gene for the protein of interest is isolated and cloned into an appropriate host for production.

- Microbes (e.g. *E. coli*) – cheap but lacks cellular machinery to produce complex proteins.
- Mammalian cell lines (e.g. Chinese hamster ovary line) – able to perform complex post-translational modification, are widely used.
- Animals (e.g. goat) – large amounts of protein produced, for example, in the animal's milk. The recombinant DNA is introduced into the animal's germ line so offspring will also produce the desired protein.

These processes are cheap, safe, and allow for easy post-production modification (e.g. with polyethylene glycol to reduce renal clearance and maximize efficiency). This technology is now used to produce a wide range of proteins, including insulin, growth hormone and blood clotting factors.

Monoclonal antibodies

Monoclonal antibodies (mAbs) are monospecific antibodies which can be raised against virtually any antigen, and will selectively bind to it *in vivo*. Monoclonal antibodies form the basis of many diagnostic tests (e.g. over the counter pregnancy tests). Therapeutically, they may act by direct neutralization, but more commonly evoke a local immune response or are combined with drugs for delivery to a target tissue (immunoconjugates – 'magic bullets').

Production of mAbs involves harvest of antibody producing B-lymphocytes from mice sensitized to the target antigen, and fusion with a tumour lymphocyte, producing an immortalized antibody-producing cell

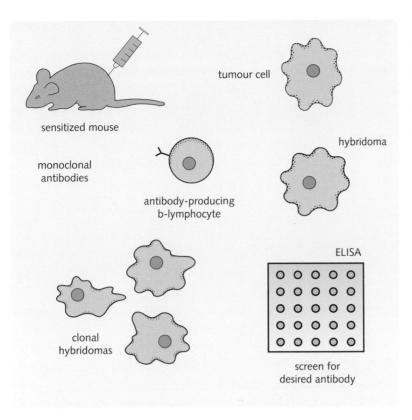

tumour cell

sensitized mouse

monoclonal
antibodies

antibody-producing
b-lymphocyte

hybridoma

clonal
hybridomas

ELISA

screen for
desired antibody

Fig. 7.21 Monoclonal antibody production. A mouse is immunized with the target antigen and antibody producing B-lymphocytes are isolated. Polyethylene glycol is used to fuse these with B lymphocyte tumour cells to create hybridomas, which are cultured on a hypoxanthine-aminopterin-thymidine (HAT) medium. Elisa is used to screen for cell clones producing the desired antibody, which are expanded. Antibodies are isolated from the medium and purified.

(a hybridoma; Fig. 7.21). Recent technologies such as phage and mammalian cell display have bypassed the need for hybridomas. To eliminate the host immune response against the murine components, humanized monoclonal antibodies are engineered and expressed in mammalian cells.

Most mAbs currently licensed are targeted against host tissues for the treatment of cancer or autoimmune diseases. However, a large number of mAbs raised against foreign pathogens are also in development. As a rule of thumb, the class can be recognized as a drug name ending '-imab' or '-umab'.

By the end of this chapter you should be able to:
- Understand the common processes of mutagenesis, and appreciate how different classes of mutation yield different effects on protein structure and function.
- Draw out example pedigrees representing autosomal dominant, autosomal recessive, X-linked dominant, X-linked recessive, holandric and mitochondrial inheritance.
- Understand the concept of consanguinity and its effect on the incidence of autosomal recessive conditions.
- Appreciate how skewed Lyonization can result in X-linked recessive disorders appearing in female patients.
- Define anticipation and understand its relevance in trinucleotide repeat diseases.
- Appreciate the concept of heritability in the context of complex diseases and describe the threshold model of multifactorial disorders.
- Understand the importance of heterozygous cancer predisposition and describe features suggestive of inherited cancer susceptibility.
- Define aneuploidy, triploidy, trisomy and monosomy, with examples of resultant diseases.
- Appreciate important current and future potential treatment options for genetic diseases.

Genetic disease is a term that encompasses not only single-gene disorders (see p. 130) and chromosomal defects (see p. 142), but also complex, multifactoral diseases (see p. 137). Although each genetic disorder may be rare, combined together genetic disease is common, and can have a major impact on morbidity and mortality. An understanding of genetics is important for the diagnosis and management of such disorders (discussed throughout this chapter), and for the identification of genetic disease 'carriers' for genetic counselling (see Ch. 9).

GENOTYPE AND PHENOTYPE

The genotype of an organism describes the information contained in the genome. The phenotype is the outward appearance of an organism for a given characteristic, and results from a complex set of interactions between the genotype and external environment.

Traits are observed variations within a specific character of the phenotype, and are coded for by alleles. A single trait may be caused by multiple different genotypes, and all of an organism's traits combined make up its phenotype.

MUTATION AND MUTATIONS

Mutations are permanent changes in the amount or structure of genetic material. They can be inherited or occur spontaneously, and can be subdivided into germline or somatic. Most genetic diseases are inherited, whereas most cancers arise from somatic mutations. For new mutations to be heritable they must occur in the germline.

The vast majority of mutations, however, do not occur in genes, and those that do rarely have pathogenic potential.

Mechanisms of mutation

At the single-gene level mutations may result from:
- substitution (point mutation)
- deletion
- insertion
- inversion
- triplet repeat expansion (unstable expansion).

Substitution

Substitution is the replacement of a single nucleotide by another with no net gain or loss of chromosomal material. Point mutations may arise as a result of mistakes in DNA replication, repair, or (most commonly) as the result of the spontaneous deamination of methylated cytosine to thymine. Substitutions are classified as:

- transition – purine to purine or pyrimidine to pyrimidine
- transversion – purine to pyrimidine or vice versa (Fig. 8.1).

Point mutations may be silent or deleterious depending upon their type and site (see p. 129). Rarely, a mutation may be advantageous and favoured by natural selection.

Deletion and insertion

Deletion is loss of DNA involving from one to many thousands of base pairs. Sequences at the ends of deletions are often similar, predisposing to recombination errors (Fig. 8.2A).

Insertion is a gain of DNA. Small duplications (a type of insertion) can occur when runs of bases and repeated motifs predispose to duplication by replication slippage (Fig. 8.2B). Large duplications arise by aberrant recombination.

The effects on the protein of deletion and insertion depend on:

- the amount of material lost or gained
- whether the reading frame is affected (see Fig. 8.3E).

Inversion

Inversions may involve anything from two to many thousands of base pairs. They typically occur in areas of sequence homology (sequences at each end of the inverted segment often resemble each other).

Triplet repeat expansions

Triplet or trinucleotide repeat expansions are a subset of unstable microsatellite repeats (see p. 126), typically involving cytosine/guanine-rich trinucleotides (CGG, CCG, CAG, CTG), that occur throughout the genome. They can involve a few copies to several thousand repeats. Expansion with each successive generation leads to 'anticipation' (see p. 135).

> **Clinical Note**
>
> In Friedreich ataxia the most common abnormality (FRDA1) is an expansion of the trinucleotide sequence GAA within the first intron of the *FXN* gene. The larger the number of GAA copies, the earlier the onset of the disease and the quicker the decline of the patient.

Structural effects of mutation on protein

Silent mutations

Silent mutations are point mutations which have no effect on the amino-acid sequence of a protein (Fig. 8.3A, B). They have long been considered to be 'evolutionary neutral', but recent studies have demonstrated that silent mutations may exert control over differential splicing.

Missense mutations

Missense mutations arise when a codon is altered, leading to the incorporation of a different amino acid into the peptide chain (Fig. 8.3A, C). The functional effects depend upon the location of the alteration relevant to the tertiary or quaternary structure of the protein, and whether the two amino acids have similar chemical properties (i.e. hydrophobic to hydrophilic is more disruptive than hydrophobic to hydrophobic).

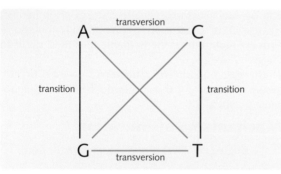

Fig. 8.1 The relationship between transition and transversion substitution mutations.

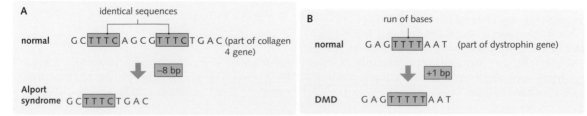

Fig. 8.2 (A) Deletion in Alport syndrome. (B) Duplication in Duchenne muscular dystrophy (DMD). bp, base pairs.

A	**Wild Type**	T A C A A C G T C A C C A T T	anti-sensa DNA
		A U G U U G C A G U G G U A A	mRNA
		↓	
		met - leu - gln - trp - stop	Protein coded
B	**Silent mutation**	T A C A A T̄ G T C A C C A T T	anti-sensa DNA
		A U G U U A C A G U G G U A A	mRNA
		↓	
		met - leu - gln - trp - stop	Protein coded
C	**Missense mutation**	T A C A A Ḡ G T C A C C A T T	anti-sensa DNA
		A U G U U C C A G U G G U A A	mRNA
		↓	
		met - phe - gln - trp - stop	Protein coded
D	**Nonsense mutation***	T A C A A C G T C A C T̄ A T T	anti-sensa DNA
		A U G U U G C A G U G A U A A	mRNA
		↓	
		met - leu - gln - stop	Protein coded
E	**Frameshift mutation***	T A C T̄ A A C G T C A C C A T T	anti-sensa DNA
		A U G A U U G C A G U G G U A A	mRNA
		↓	
		met - ile - ala - val - val -	Protein coded

Fig. 8.3 The structural effects of mutation on protein. (A) Wild type. (B) Silent mutation. (C) Missense mutation. (D) Nonsense mutation. (E) Frameshift mutation. * nonsense and frameshift RNA is usually degraded, so no protein is produced. A squared base indicates a change from the wild-type.

Clinical Note

In sickle-cell disease, the substitution of A by T at the 17th nucleotide of the β-globin gene changes the codon for the 6th amino acid from GAG (for glutamic acid) to GTG (which encodes valine). This missense mutation changes the solubility and molecular stability of the resultant haemoglobin causing the formation of polymers, leading to sickling of red blood cells.

Nonsense mutations

Nonsense mutations lead to the conversion of a codon to a stop codon (UAG, UAA, UGA) (Fig. 8.3A, D). The nonsense mRNA is mostly subjected to degradation and no protein product is formed. If the nonsense mutation occurs close to the end of the sequence a truncated protein may be formed, but these products are rarely functional.

Frameshift mutations

Insertions and deletions of nucleotides, if not a multiple of three, lead to 'frameshift' mutations whereby the open reading frame of the gene and the corresponding amino-acid sequence is altered (Fig. 8.3A, E). The corresponding abnormal protein is rarely translated as the frame-shifted RNA transcripts are usually degraded.

Functional effects of mutation on protein

With the exception of imprinted genes (see p. 135), genes on both the maternal and paternal chromosomes are expressed. If either gene contains a mutation, the cell may express two different protein products. Mutations exert their phenotypic effects by one of two mechanisms: loss of function or gain of function.

Loss of function mutations
Null mutation
Null mutations, also known as 'amorphic mutations', are associated with a complete absence of gene product function.

Hypomorphic mutation
Hypomorphic mutations, also known as 'leaky mutations', lead to a partial loss of function. They usually result from:

- an altered amino acid that makes the polypeptide less active
- a reduction in transcription that results in less normal transcript.

They are usually recessive to wild type.

Haploinsufficiency
The majority of heterozygous states are haplosufficient; that is one functional copy of a gene is adequate for the manifestation of a wild type phenotype. Haploinsufficiency describes a situation whereby a reduction of 50% of gene function results in an abnormal phenotype. Such mutations are invariably dominant.

Gain of function mutations
These mutations result in either:

- increased activity of the gene product (hypermorphic)

or

- the gain of a novel function or a novel pattern of gene expression of the gene product (neomorphic).

They are usually dominant.
   Trinucleotide repeat expansions also represent gain of function mutations, usually a toxic gain of protein function, which predisposes to protein misfolding and aggregation, leading to neurodegeneration.

Dominant negative mutations

Dominant negative (antimorphic) mutations arise when the null allele product of a heterozygote adversely affects the normal gene product, for example by dimerizing with and inactivating it. Classically the abnormal protein is unable to function in a multimeric complex, as seen with fibrillin in Marfan syndrome.

MONOGENIC DISORDERS

Monogenic disorders are caused by individual mutant genes, and frequently show characteristic patterns of inheritance. There are approximately 6000 monogenic disorders and although individually they are rare, together they affect approximately 1% of the population. Some need an environmental trigger before the phenotype is expressed. For example, severe emphysema in individuals homozygous for α_1-antitrypsin deficiency mutations is largely confined to smokers. Single-gene disorders generally follow Mendelian patterns of inheritance.

An introduction to pedigrees

A pedigree is a record of an individual's ancestors, offspring, siblings and their offspring that can be used to determine the pattern of certain genes or disease inheritance within a family (Fig. 8.4).

Mendelian inheritance of single gene disorders

Mendelian traits generally occur in predictable proportions among the offspring of parents with that trait. The pattern of inheritance seen depends on the chromosomal location of the gene (sex-linked or autosomal) and whether the phenotype is dominant or recessive. Therefore, there are five patterns of Mendelian single-gene inheritance:

1. Autosomal dominant
2. Autosomal recessive
3. X-linked dominant
4. X-linked recessive
5. Y-linked/holandric.

A sixth pattern of single-gene inheritance, mitochondrial, is non-Mendelian (see p. 136).

'Dominant' phenotypes are expressed in heterozygotes, whereas 'recessive' traits are expressed only in homozygotes. If the expression of each allele can be detected in the presence of the other, the alleles are 'codominant'.

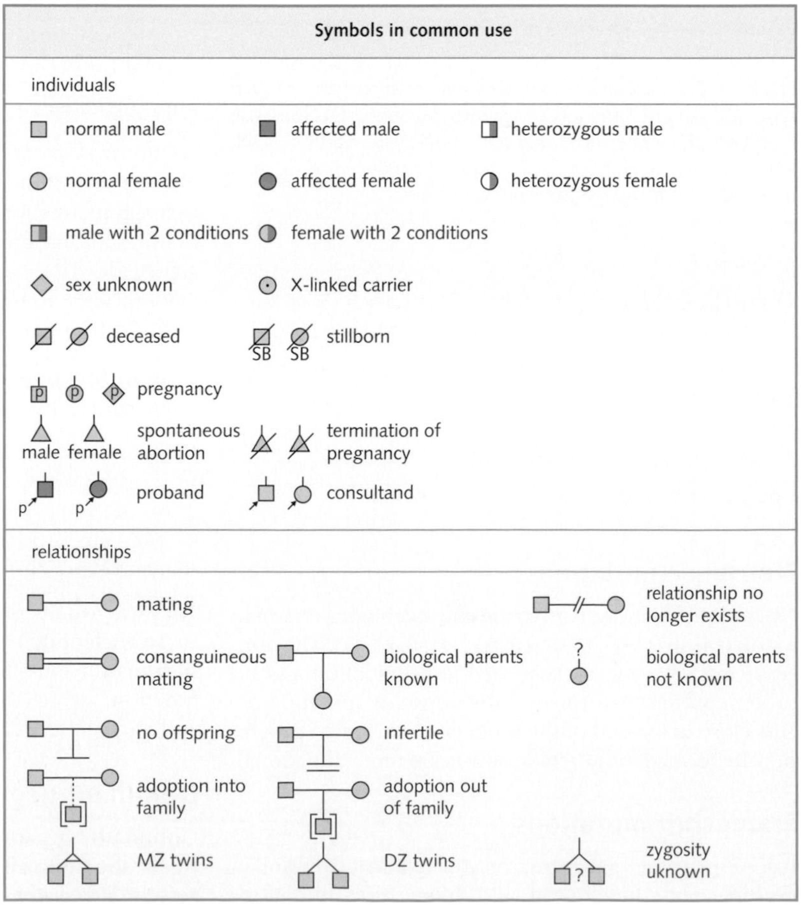

Fig. 8.4 Symbols and configuration of pedigree charts.

Symbols in common use

individuals

□ normal male	■ affected male	▣ heterozygous male
○ normal female	● affected female	◖ heterozygous female
▤ male with 2 conditions	◓ female with 2 conditions	
◇ sex unknown	⊙ X-linked carrier	
⊘ ⊘ deceased	⊘SB ⊘SB stillborn	
℗ ℗ ◈ pregnancy		
△ △ spontaneous male female abortion	⋇ ⋇ termination of pregnancy	
⤢ proband	⤢ consultand	

relationships

□—○ mating		□—//—○ relationship no longer exists
□══○ consanguineous mating	□┬○ biological parents known	? biological parents not known
□┐┌○ no offspring	□‖○ infertile	
□—○ adoption into family	□—○ adoption out of family	
MZ twins	DZ twins	zygosity unknown

Dominant and recessive inheritance is defined according to clinical phenotypes, which may not always reflect the behaviour of the allele at the molecular level. Mutations in the retinoblastoma protein are recessive at the cellular level because one allele expresses enough protein for biological function. However, at the phenotypic level the predisposition to cancer is inherited dominantly (see also p. 139).

Autosomal dominant disorders

Autosomal dominant inheritance

A dominant allele is phenotypically expressed in homozygotes and heterozygotes (Fig. 8.5).

- Affected parents have affected children – vertical inheritance.
- Unaffected family members usually have unaffected partners and they produce unaffected children (assuming the condition does not arise *de novo*).
- Affected family members usually have unaffected partners and there is therefore a 50% risk to each child of being affected (assuming full penetrance – see p. 135).
- Usually both sexes are equally affected.

Homozygotes for the trait are rare. In some autosomal dominant (AD) conditions, new mutations account for a substantial proportion of cases (e.g. achondroplasia, familial adenomatous polyposis). AD genes can show:

- sex limitation (e.g. testicular feminization)
- reduced penetrance (e.g. retinoblastoma)
- variable expressivity (e.g. tuberous sclerosis)
- imprinting (see p. 135)
- anticipation (see p. 135).

example pedigree

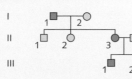

mating of affected parent (Aa)

		affected parent	
gametes		A	a
unaffected parent	a	Aa	aa
	a	Aa	aa

1:2 risk to each pregnancy of producing an affected child (i.e. Aa as shown)

Fig. 8.5 Example pedigree and typical offspring of mating in autosomal dominant inheritance. A, disease allele.

Molecular basis of autosomal dominant inheritance

In AD disorders, the disease occurs despite the presence of the normal gene product expressed from the wild type allele. This may arise as a result of:

- haploinsufficiency – e.g. familial hypercholesterolaemia (Fig. 8.6)
- dominant negative effect – e.g. osteogenesis imperfecta (Fig. 8.7)
- loss of heterozygosity – loss of normal function of the wild-type allele due to somatic mutation (e.g. of tumour suppressor genes – see p. 141)
- simple gain of function – from increased expression of the normal protein, e.g. Charcot–Marie–Tooth disease type 1a; or expression of an abnormal protein with novel properties, e.g. Huntington disease.

> **Clinical Note**
>
> Familial hypercholesterolaemia is an AD disorder, resulting from mutation in the LDL receptor gene on chromosome 19, with an incidence of 1:500.

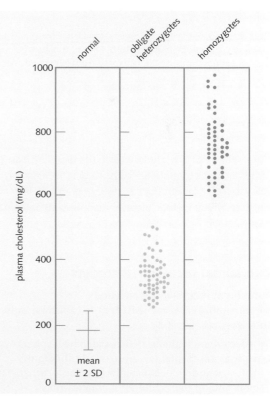

Fig. 8.6 Gene dosage in familial hypercholesterolaemia. Heterozygotes show levels of plasma cholesterol intermediate between normal and homozygotes for the mutation. The plasma cholesterol level in heterozygotes is sufficient for the development of premature heart disease, so the condition shows an AD pattern of inheritance. (5 mmol/L is approx. 193 mg/dL). (Adapted from Nussbaum, McInnes and Willard, 2001.)

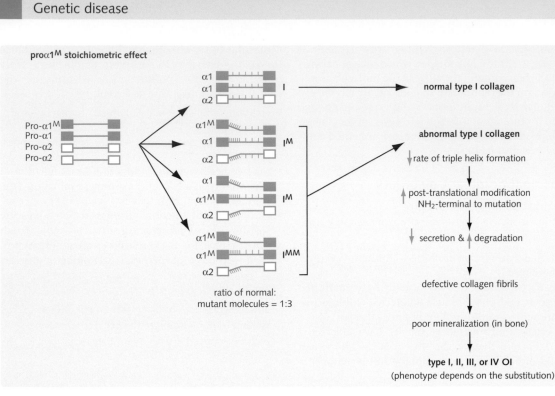

Fig. 8.7 Dominant negative effect. Normal collagen consists of two molecules of pro-α1 and one molecule of pro-α2. Procollagen containing a missense mutation (pro-α1^M) destabilizes the collagen triple helix, resulting in increased degradation, which results in OI, osteogenesis imperfecta. (Adapted from Nussbaum, McInnes and Willard, 2001.)

The half-life of LDL cholesterol in the blood increases from 2.5 days to around 5 days, leading to markedly elevated LDL levels. The sequelae of high LDL levels include atherosclerotic disease, tendon xanthomata, xanthelasma and premature corneal arcus.

Autosomal recessive disorders

Autosomal recessive inheritance

An autosomal recessive (AR) trait is expressed only in homozygotes (Fig. 8.8).

- Affected individuals will usually have phenotypically normal parents – horizontal inheritance.
- Affected individuals usually have non-carrier partners and all their children will be carriers.
- If a carrier has an non-carrier partner, there is a 50% chance of the children being carriers.
- Only matings between heterozygotes will produce affected individuals, with an expected frequency of 1 in 4.
- Both sexes are equally affected.
- There is an association with consanguinity due to sharing of alleles in families.

mating of two carriers (Aa)

gametes	carrier	
	A	a
carrier A	AA	Aa
a	Aa	aa

25% affected, 50% carriers, 25% unaffected non-carriers

mating of carrier and unaffected

gametes	carrier	
	A	a
unaffected A	AA	Aa
A	AA	Aa

50% carriers, all phenotypically unaffected

Fig. 8.8 Example pedigree and typical offspring of matings in autosomal recessive inheritance. a, disease allele.

Consanguinity is mating between two people who have a familial relationship closer than that of second cousins.

Complementation

Complementation is the ability of two different genetic defects to correct for one another. In recessive conditions it is expected that two affected parents will always have affected children. However, occasionally two such parents will have an unaffected child as a result of complementation. Complementation arises if parents who have a similar clinical phenotype are homozygous for mutations in different genes in the same pathway (Fig. 8.9).

Molecular basis of recessive inheritance

AR disorders include many enzyme defects where expression from the wild-type allele in the heterozygote provides sufficient functional protein to prevent disease. However, homozygotes express no functional protein and develop the disorder.

Examples of AR conditions include:

- cystic fibrosis
- phenylketonuria
- haemochromatosis
- Gaucher disease.

Clinical Note

Phenylketonuria (PKU) is an AR disease, resulting from mutation of phenylalanine hydroxylase encoded on chromosome 12, and the inability to convert phenylalanine into tyrosine leading to the accumulation of phenylalanine and its metabolic products in body fluid. If unrecognized and untreated, PKU leads to irreversible mental disability, microencephaly and fits. PKU is controllable with dietary phenylalanine restriction.

X-linked disorders

X-linked dominant inheritance

The X-linked dominant (XD) inheritance pattern is rare and difficult to distinguish from AD except that affected males have normal sons, but all daughters are affected (Fig. 8.10).

Examples include:

- Aicardi syndrome
- vitamin D resistant rickets
- Rett syndrome.

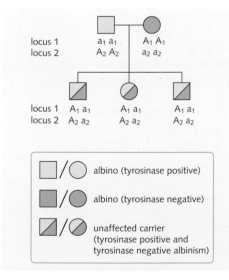

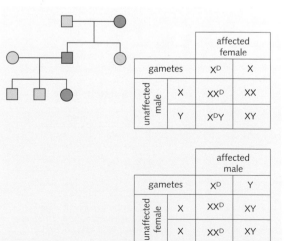

Fig. 8.9 Pedigree showing complementation. Both parents are albino, but all their children are normal. Albinism is an autosomal recessive condition, but the parents are homozygous for mutations in different genes in the same pathway. The children are unaffected carriers of both mutations. A_1, wild-type at locus 1, a_1, mutant at locus 1, A_2, wild-type at locus 2, a_2, mutant at locus 2.

Fig. 8.10 Example pedigree and typical offspring of matings in X-linked dominant inheritance. X^D, disease allele.

X-linked recessive inheritance

For X-linked recessive (XR) genes the inheritance pattern (Fig. 8.11) is as follows:

- many more males than females show the phenotype
- the disease is transmitted by a carrier female, who is usually asymptomatic
- if a mother is a carrier, her sons have a 50% chance of being affected and her daughters a 50% chance of being carriers
- an affected male will usually have no affected offspring, but all his daughters will be carriers
- no sons of the affected male will inherit the mutant allele (i.e. no male-to-male transmission)
- affected males may have unaffected parents, but they may have an affected maternal uncle or male cousin.

Molecular basis of X-linked recessive inheritance

Males (XY) have only one X chromosome and are therefore said to be hemizygous for X-linked genes. Since males receive only one copy of X-linked genes (except for those in the pseudo-autosomal region) they will express any XR traits because they have no compensating wild-type allele.

To compensate for the double complement in females, X chromosome inactivation (lyonization) ensures that genes are only expressed from one of the two X chromosomes. The selection of an X chromosome for inactivation within a specific cell is random, but once established early in development inactivation patterns are transmitted to daughter cells. Females do not tend to show XR disease because many of their cells express the wild-type allele. Women can be affected with X-linked recessive conditions in the following situations:

- if she is the daughter of an affected male and a carrier female
- if there is skewed lyonization of a 'non-diseased' X chromosome
- if there is X chromosome-autosome translocation
- if XO (Turner syndrome) is present.

XR conditions include:

- G6PDH deficiency
- haemophilia A and B
- Duchenne muscular dystrophy
- Fabry disease.

Y-linked (holandric) inheritance

As only males have the Y chromosome, only males can pass on the Y to offspring, and only male offspring can receive it. In order to allow it to pair with the X chromosome at cell division; the Y chromosome contains pseudoautosomal regions.

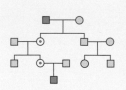

mating of affected male

		affected male	
	gametes	X^D	Y
unaffected female	X	X^DX	XY
	X	X^DX	XY

all daughters carriers, all sons unaffected non-carriers

mating of carrier female

		unaffected male	
	gametes	X	Y
carrier female	X^D	X^DX	X^DY
	X	XX	XY

half of children inherit gene regardless of sex, 50% daughters carriers, 50% sons affected, 50% children unaffected non-carriers

Fig. 8.11 Example pedigree and typical offspring of matings in X-linked recessive inheritance. X^D, disease allele.

Heterogenicity in Mendelian disorders

Genetic heterogenicity

This is the phenomenon by which identical or similar phenotypes arise by different genetic mechanisms, including different allelic mutations in the same gene, or mutations in different genes. Such disorders may show more than one mode of inheritance; e.g. Charcot–Marie–Tooth disease is usually inherited in an autosomal dominant manner, but autosomal recessive and X-linked variants exist.

Allelic heterogenicity

Different mutations in the same gene may result in the same or different phenotypes. This is an important cause of clinical variation, for example, specific mutations within the *CFTR* gene are associated with pancreatic sufficient as opposed to pancreatic insufficient forms of cystic fibrosis.

Locus heterogenicity

This is the situation in which mutations at two or more distinct loci can produce the same or similar phenotype. For example, retinitis pigmentosa may result from mutations in many different genes.

Complications to Mendelian pedigree patterns

A number of disorders have been identified that do not follow classic patterns of Mendelian inheritance. The molecular mechanisms underlying these observations are now beginning to be understood.

Mechanisms disguising basic Mendelian pedigrees

Penetrance and expressivity

The phenotype expressed is the result of a complex set of interactions between the associated genotype, other genes, and the environment. Therefore, for some traits the phenotype might not occur as often as the genotype due to the interference of protein expression by, for example, other gene products (epigenetic factors).

Penetrance is the proportion of individuals with a given genotype, which show the associated phenotype. Incomplete penetrance means that monogenetic diseases do not always have predictable expression patterns in a population.

Expressivity is the degree to which a trait is expressed. Unlike penetrance, expressivity describes individual variability rather than statistical variability in a population.

Anticipation

Anticipation is the occurrence of a hereditary disease with a progressively earlier age of onset in successive generations. Mutations associated with anticipation are trinucleotide repeat expansions, which often grow in successive generations (e.g. Huntington disease and myotonic dystrophy). Larger expansions are associated with earlier age of onset.

Imprinting

This is differential expression of genetic material depending upon which parent it has been inherited from. It is thought to result from the selective inactivation of genes in different patterns in male and female gametogenesis. Hydatidiform moles illustrate the different roles of paternal and maternal genomes:

- A complete mole (46,XX) has chromosomes that are all paternal in origin (both X chromosomes being of paternal origin, i.e. extra paternal set, but no maternal set), and results in either no foetus, or a normal placenta with severe hyperplasia of the cytotrophoblast.
- A partial mole (69,XXX or 69,XXY) is triploid with an extra set of chromosomes of maternal or paternal origin. An extra paternal set (diandric) results in abundant trophoblast, but poor embryonic development. An extra maternal set (digynic) results in severely retarded embryonic development with a small fibrotic placenta.

Imprinting is important in the aetiology of Prader–Willi (PW) and Angelman syndromes (AS), which arise from the same microdeletion of chromosome 15q11–13 (the 'critical region'). The phenotype varies between PW and AS according to whether the deleted critical region is paternally or maternally inherited.

AS results from imprinting being switched off in the critical region on the paternally derived chromosome, and PW from the same on the maternally derived chromosome (Fig. 8.12). When a child inherits a chromosome 15 with the microdeletion:

- AS results from the loss of the imprinted maternal locus, thus removing the only active copy of the responsible gene, *UBE3A*
- PW syndrome results from the loss of the imprinted paternal locus, removing the active copies of the several genes postulated as having a role in PW.

Uniparental disomy

This is duplication of a chromosome from one parent, with loss of the corresponding homologue from the other (Fig. 8.13). For example, uniparental disomy of maternal chromosome 15 can result in the same phenotype as PW, but with no deletion because there is no paternally contributed chromosome 15. Beckwith–Wiedemann syndrome can be due to paternal duplication of 11p15.

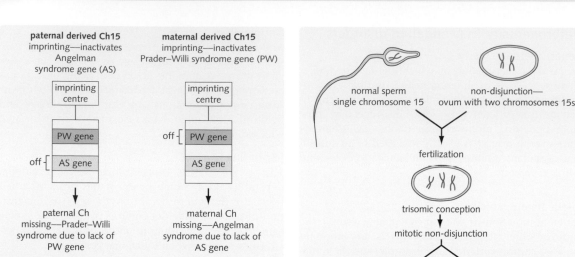

Fig. 8.12 Deletion of one parental chromosome 15 with imprinting of another can lead to Prader–Willi (PW) or Angelman syndrome (AS). Imprinting of the PW and AS genes is under the control of the imprinting centre.

Mitochondrial inheritance

Mitochondrial DNA (MtDNA) is only maternally inherited, so affected males cannot transmit the disease to their offspring (Fig. 8.14). Mitochondria are distributed randomly in daughter cells, so these may contain entirely normal mitochondrial DNA or entirely mutant DNA (homoplasmy), or a mixture of both (heteroplasmy), leading to variable expression of disease depending upon the relative proportion of normal to mutant DNA. Mitochondrial diseases include:

- Leber hereditary optic neuropathy
- mitochondrial encephalomyopathy, lactic acidosis and stroke-like syndrome (MELAS)
- myoclonus with epilepsy and with ragged red fibres (MERRF).

Mosaicism

A mosaic is an individual with multiple cell lines (which exhibit different genotypes) that arise from a single zygote. Germline mosaicism occurs when an abnormal cell line is confined to the gonads, and it may account for apparently unaffected parents producing more than one child with an AD condition. Somatic mosaicism usually occurs due to mitotic errors in early embryonic cleavage, causing different cell lines to develop.

Mosaicism can be an important clinical factor in some instances of chromosomal disorders. Down syndrome (DS) may be mosaic (46,XY/47,X+21 or 46,XX/47,XX+21)) (Fig. 8.15) as may Klinefelter and Turner syndrome (see p. 146).

A rare form of somatic mosaicism is chimerism, the fusion of two or more fertilized zygotes in early embryological development. This accounts for no important examples of disease in humans.

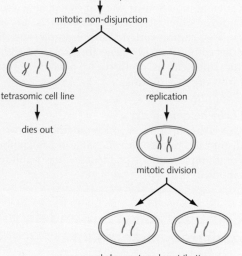

Fig. 8.13 Mechanism of uniparental disomy (for chromosome 15).

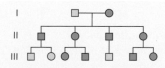

Fig. 8.14 Example pedigree in mitochondrial inheritance. (Adapted with permission from Turnpenny & Ellard Emery's Elements of Medical Genetics 12 e, Churchill Livingstone, 2005.)

> ### HINTS AND TIPS
>
> All females are genetic mosaics for the X chromosome. As X chromosomal inactivation is a random process, in some cells the maternally derived X chromosome will be lyonized to yield a Barr body, while in others the paternally inherited copy is 'switched off'.

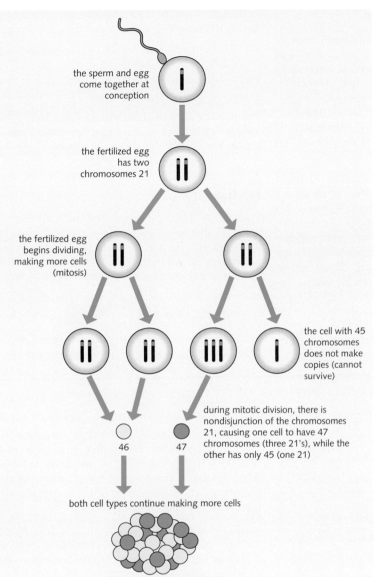

Fig. 8.15 Genetic mosaicism. Mitotic non-disjunction leads to two distinct populations of cells. Severity of any resultant condition is linked to the proportion of mutated cells to normal cells. The earlier in development the non-disjunction event occurs, the higher the ratio of mutated cells.

the sperm and egg come together at conception

the fertilized egg has two chromosomes 21

the fertilized egg begins dividing, making more cells (mitosis)

the cell with 45 chromosomes does not make copies (cannot survive)

during mitotic division, there is nondisjunction of the chromosomes 21, causing one cell to have 47 chromosomes (three 21's), while the other has only 45 (one 21)

46   47

both cell types continue making more cells

this results in mosaicism (two types of cells) for trisomy 21 in the developing baby

POLYGENETIC INHERITANCE AND MULTIFACTORIAL DISORDERS

Multifactorial disorders typically result from mutations in several genes, in combination with environmental factors. Although multifactorial disorders tend to recur in families, they are much more prevalent than single-gene disorders and do not show Mendelian inheritance patterns. Multifactorial inheritance is implicated in many common conditions including:

- congenital malformations – e.g. neural tube defects, developmental dysplasia of the hip (DDH), pyloric stenosis, cleft lip and palate and congenital heart disease
- common disorders of adult life – e.g. diabetes mellitus, obesity, epilepsy, hypertension and schizophrenia
- normal human characteristics – e.g. blood pressure, height, dermatoglyphics (finger ridges) and intelligence.

Multifactorial disorders

A continuous normal (Gaussian) distribution curve of the trait within the population as a whole is typical of multifactorial disorders. Abnormalities do not usually have a distinct phenotype, but are extremes of the curve. The number of genes involved may be very few as environmental variations can ensure normal distribution.

Twin studies highlight the relative importance of genes and the environment. For example, cleft lip and palate has a population incidence of 1 in 1000, but:

- concordance in monozygotic twins is 40% – if due to genetics alone there would be 100% concordance
- concordance in dizygotic twins is 4% – these twins are not genetically identical (having on average, like all siblings, 50% of their genes in common), but they generally share a similar environment, showing the importance of environmental factors where the concordance rate=[both affected / (one affected + both affected)] × 100.

Threshold model of multifactorial disorders

Figure 8.16 shows the threshold model of multifactorial disorders. The liability of a population to a particular disease follows a normal distribution, with an individual's genetic susceptibility and environmental factors defining their liability.

- The disorder is manifested when a certain threshold of liability is exceeded.
- Population incidence is equivalent to the proportion whose liability is greater than the threshold.

The liability curve is shifted to the right for relatives of an affected individual, such that more members of the family are likely to be above the disease threshold.

Heritability

Heritability is the proportion of the total phenotypic variation that is genetic in origin in a given population, expressed as a percentage or decimal. For example, the heritability of schizophrenia is 85% (0.85). Those of asthma and rheumatoid arthritis are 60% (0.60).

> **HINTS AND TIPS**
>
> Heritability estimates are population specific since the variation of environmental and genetic effects may not be identical in different geographical areas and ethnic populations.

If heritability is high, there is a high correlation in relatives (e.g. finger ridge correlation in first-degree relatives is 49%). Usually heritability is low, so the incidence in relatives falls off sharply (Fig. 8.17). Recurrence risks in multifactorial disorders are based upon population and family studies and are influenced by:

- the severity of the disorder in the affected person (the 'proband') in some conditions (e.g. severe bilateral cleft lip is more likely to recur in siblings than unilateral cleft lip)
- the number of affected individuals in a family
- the proband being of the less commonly affected sex

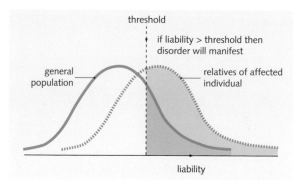

Fig. 8.16 Threshold model of multifactorial inheritance.

Fig. 8.17 Risk to relatives for multifactorial disorders. First-degree relatives are parents, siblings and offspring (share 50% of genome); second-degree relatives are grandparents, aunts and uncles, grandchildren (share 25% of genome); third-degree relatives include cousins and great-grandchildren (share 12.5% of genome).

Disorder	Relative risk disorder			
	1° relative	2° relative	3° relative	General population
Cleft lip and palate	4	0.6	0.3	0.1
Neural tube defects	4	1.5	0.6	0.3
Epilepsy	5	2.5	1.5	1.0

• the relationship to the proband – the recurrence risk for first-degree relatives is approximately the square root of the population incidence of the trait.

There may be a sex difference in population incidence. For example, pyloric stenosis is more common in boys, so children of an affected female will be more likely to develop the condition than children of an affected male, as the female needs a high number of risk genes to manifest the condition.

Environmental factors

The judicious manipulation of environmental factors may enable the reduction in an individual's susceptibility to below the disease threshold value. For example, atherosclerosis has a heritability of 65%, so environmental contribution is approximately 35%. Epidemiological studies show that not smoking, healthy eating and taking regular exercise significantly reduce an individual's risk of developing heart disease.

GENETICS OF CANCER

In the majority of cases, cancer is a multifactorial disorder. However, in a minority (about 5%) the disease follows a familial pattern of transmission, suggesting a genetic cancer syndrome. Characterization of the mutations segregating in such families has helped elucidate the molecular events that underlie the more common multifactorial form of the disease.

Normal cell proliferation and survival is controlled by growth promoting proto-oncogenes and growth inhibiting tumour suppressor genes (see Ch. 6). Mutation in these genes may result in cancer (Figs 8.18, 8.19).

Multistage process of carcinogenesis

Carcinogenesis requires the accumulation of many mutations in both oncogenes and tumour suppressor genes. Tumours evolve from benign to malignant as subpopulations of cells acquire further mutations (Figs 8.20, 8.21). Given that mutations may arise in any tumour cell, tumours tend to be mosaics. However, when a cell acquires a mutation associated with a further loss of growth inhibition, this cell type will dominate.

Oncogenesis

Oncogenes drive cell growth (see Ch. 6); they are usually dominant at the cellular level, with most mutations being gain of function mutations. They can be classified

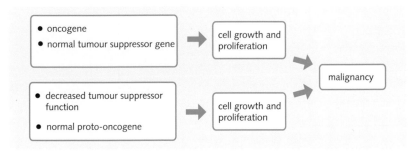

Fig. 8.18 Pathways to malignancy.

Fig. 8.19 Differences between oncogenes and tumour suppressor genes.	
Oncogene	**Tumour suppressor gene**
Gene active in tumour	Gene inactive in tumour
Specific translocations/point mutations	Deletions or mutations
Mutations rarely hereditary	Mutations can be inherited
Dominant at cell level	Recessive at cell level
Broad tissue specificity	Considerable tumour specificity
Especially leukaemia and lymphoma	Solid tumours

139

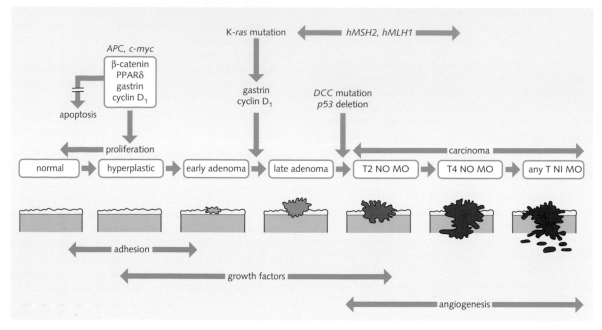

Fig. 8.20 The stages in the evolution of colorectal cancer (the adenoma–carcinoma sequence). (Adapted from Kumar and Clark, 2005.)

Fig. 8.21 The multistep model of carcinogenesis.

Phase	Mechanism	Clinical appearance
Initiation	Mutation	No noticable change
Promotion	Mutated cells stimulated to grow	Usually only detected histologically/biopsy
Conversion	Mutation: Uncontrolled growth and expansion	Benign tumour, expansive growth
Progression	Mutation: Complete loss of cellular control	Cancer: Invasion and metastasis

Fig. 8.22 Classes of oncogene by transduction position. (1) Growth factors (e.g. platelet-derived growth factor (PDGF)-*sis*). (2) Growth factor receptors (e.g. EGFR-*Gb*). (3) Post-receptor proteins (e.g. *ras*). (4) Post-receptor tyrosine kinase (e.g. *abl*, *src*). (5) Cytoplasmic proteins (e.g. *raf*). (6) Nuclear proteins (e.g. *myc*).

according to their position in the normal signal transduction pathway (Fig. 8.22).

Activation of oncogenes may occur by:

- translocation (e.g. Burkitt lymphoma t(8;14) or t(8;2) activates c-*myc* on 8 q
- amplification (e.g. n-*myc* amplified in neuroblastoma)
- point mutation in an oncogene (e.g. Ha-*ras* mutation in bladder cancer).

HINTS AND TIPS

Proto-oncogenes are genes that have the potential to mutate into oncogenes.

Tumour suppressor genes

Tumour suppressor genes (Fig. 8.23) inhibit oncogenesis (see Ch. 6). Tumours develop if there is loss of both wild-type alleles (Fig. 8.24). Loss of activity can occur through damage to the genome (e.g. mutation, rearrangement or mitotic recombination), or interaction with

Fig. 8.23 Examples of tumour suppressor genes.

Gene	Locus	Function	Tumour
Rb	13q14	Substrate of CDK (cell cycle regulation)	Retinoblastoma, osteosarcoma
p53	17p13	Growth arrest and apoptosis	Mutated in 70% of all human tumours
WT-1	11p13	Zinc finger	Wilms' tumour
NF-1	17q	Transcription factor	Neurofibromatosis
APC	5q21	Cell adhesion	Adenomatous polyposis
DCC	18q21	Cell adhesion	Colorectal cancer, pancreatic cancer, oesophageal cancer

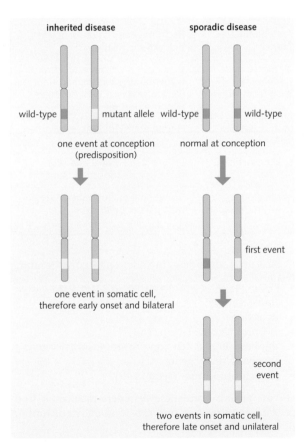

Fig. 8.24 Knudson's 'two-hit' hypothesis. Note that two events must occur to lose tumour repressor function.

cellular or viral proteins (e.g. MDM2, HPV E6 antigen or adenovirus E1b protein).

Tumour suppressor genes are important in:

- inherited predisposition to cancer
- early events in tumourigenesis (co-operation with dominant transforming genes).

The detection of mutations in tumour suppressor genes can be used for presymptomatic diagnosis (e.g. of adenomatous polyposis coli by demonstrating mutations in the *APC* gene).

Tumour suppressor genes in inherited cancers

Tumour suppressor genes are recessive at the cellular level. Therefore, if an individual inherits a mutation in one allele, every cell in the body will be relying on the product of the wild-type allele. It is highly likely that the wild-type allele will undergo somatic mutation in at least one cell. Thus, inherited mutations in tumour suppressor genes, such as in retinoblastoma and Li–Fraumeni syndrome tend to be inherited dominantly. This phenomenon is called 'loss of constitutional heterozygosity' (LOCH).

Features suggestive of inherited cancer susceptibility in a family

Important features include:

- several close (first or second degree) relatives with the same type of cancer
- several close relatives with genetically associated cancers (e.g. breast and ovary, or bowel and endometrial)
- two family members with the same rare cancer
- an unusually early age of onset
- bilateral tumours in paired organs
- synchronous or successive tumours
- tumours in two different organ systems in one individual.

Genetic cancer syndromes

A number of cancers have a genetic predisposition and genetic cancer syndromes are thought to account for 5–10% of all cancer cases. They are often characterized by the features described above, and the relatives of patients suspected or identified as having a cancer

Fig. 8.25 A summary of some important genetic cancer syndromes. Mode [of inheritance]: AD, autosomal dominant; AR, autosomal recessive; MEN, multiple endocrine neoplasia syndrome.

Syndrome	Mode	Responsible gene	Associated tumours	Other features
Ataxia telangiectasia	AR	ATM (11q22.3)	Lymphatic leukaemia, others	Childhood onset, cerebellar ataxia, telangiectasias, radiosensitivity Median age of death 20 years
Familial breast cancer	AD	BRCA1 (17q21)	Breast, ovarian, others	Presents early, accounts for 5% of all breast cancer
	AD	BRCA2 (3q12.3)	Breast, endometrial, renal, many others	Common in Ashkenazi Jews
Familial adenomatous polyposis coli	AD	APC (5q21)	Multiple benign adenomas, adenocarcinoma	Characteristic retinal changes Risk of malignancy >90%
Li–Fraumeni syndrome	AD	p53 (17p13.1)	Soft tissue and adrenal carcinomas, brain tumours	Most often presents in childhood
Hereditary non-polyposis colon cancer (Lynch syndrome)	AD	MSH2 (2p22), MLH1 (3p21)	Colorectal, endometrial, ovarian, renal	Perform yearly colonoscopy in first-degree relatives
MEN I	AD	Tumour suppressor gene (11q13)	Pituitary adenoma, pancreatic adenoma	Hyperparathyroidism, accounts for 50% of patients with Zollinger–Ellison syndrome
MEN II	AD	RET (10q11.2)	Phaeochromocytoma, thyroid carcinoma	Hyperparathyroidism, megacolon
Xeroderma pigmentosum	AR	At least 9 subtypes	Multiple skin cancers	Progressive corneal and skin scarring on exposure to sunlight
Gorlin syndrome	AD	PTCH (9q22.3)	Basal cell carcinoma, medullobastoma, ovarian fibroma	Congenital malformations (dental, cleft palate, bifid ribs)

syndrome should also be examined and tested. Some important genetic cancer syndromes are summarized in Figure 8.25.

CHROMOSOMAL DISORDERS

Chromosomal disorders simultaneously change the dosage of many genes, resulting in major phenotypic effects. Although these occur in over 7.5% of conceptions, live-birth incidence is only 6 in 1000 since most end in spontaneous abortion. Chromosomal disorders may be numerical or structural. Numerical disorders concern:

- extra single chromosomes (e.g. trisomy)
- missing single chromosomes (e.g. monosomy – lethal except for X0)
- extra haploid sets (e.g. tetraploids). Polyploidy is incompatible with life.

These disorders result in gene-dosage effects. In trisomy disease results from over-expression of genes. In monosomy disease results from haploinsufficiency.

Structural disorders include conditions resulting from:

- translocation
- inversion
- isochromosome (see p. 148)
- duplication and deletion of chromosomal segments involving many genes
- ring chromosomes (see p. 148).

Disease arising from these disorders may result from gene dosage effects, or misexpression due to disruption of regulatory regions.

Nomenclature used for chromosomal disorders

International Standard Chromosome Nomenclature (ISCN):

- numerical disorders are described by: number of chromosomes, sex chromosomes, + or − chromosome number. For example, a boy with trisomy 21 is [47, XY,+21], Turner syndrome is [45,XO]

- structural disorders are described by: number of chromosomes, sex chromosomes, mutation (chromosomes involved); break points, margins or region; p – short arm, q – long arm.

Examples of ISCN for structural disorders:

- translocation (t) – [46,XY,t(14;21)(q11;p10)]
- inversion (inv) – [46,XY,inv(9)(p12,q14)] pericentric inversion
- isochromosome (I) – [46,X,I(Xq)] long chromosome arm of X duplicated
- duplication (dup) – [46,XY,dup(5)(q20–q30)]
- deletion (del) and ring chromosome (r) – [46,XY,del (15)(q11–q13)] in Prader–Willi syndrome; [46,XX,r (X)(p12,q14)].

> **HINTS AND TIPS**
>
> Mind your 'p's and q's'! The shorter arm of a chromosome is the 'p' arm (from the French 'petit'). The long arm is 'q', because it comes after 'p' in the alphabet!

Mechanisms leading to numerical chromosomal disorders

Polyploidy

Polyploids arise as a result of:

- fertilization by two sperm
- a diploid sperm or ovum due to failure in meiosis.

Trisomies

Trisomies may result from the failure of separation ('non-disjunction') of homologous chromosomes at meiosis I, or from the failure of separation of chromatids in meiosis II (Fig. 8.26). Advancing maternal age is the most significant risk factor for all aneuploidy.

> **HINTS AND TIPS**
>
> Most cases of autosomal trisomy result from non-disjunction at meiosis I in the female germ line.

Monosomies

Monosomies may result from non-disjunction (Fig. 8.26) or from anaphase lag (delay in the movement of one chromosome from the metaphase plate during anaphase, resulting in the loss of the chromosome).

Examples of numerical chromosomal disorders

Autosomal disorders

There are only three well-defined autosomal trisomies that are compatible with postnatal survival.

Trisomy 21 – Down syndrome

Down syndrome (47,XX+21 or 47,XY+21) typically arises as a result of non-disjunction. In 5% of cases however, it arises as a result of a Robertsonian translocation

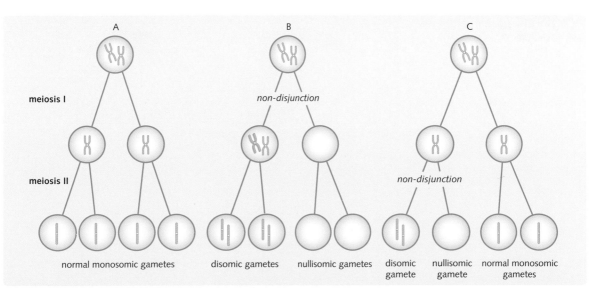

Fig. 8.26 Segregation at (male) meiosis of a single pair of chromosomes. (A) Normal meiosis. (B) Non-disjunction in meiosis I. (C) Non-disjunction in meiosis II. Female meiosis yields a single ovum and two polar bodies. (Adapted from Mueller and Young, 2001.)

(see p. 147) or of mosaicism (see p. 136). The overall incidence of DS, adjusted for the impact of antenatal screening (see Ch. 9), is 1 in 700 live births.

HINTS AND TIPS

Advancing maternal age is a significant risk factor in the aetiology of Down syndrome (DS). However, the absolute numbers of mothers having children with DS is greater in the younger age groups as the total number of mothers in these groups is higher.

Features associated with DS include:

- typical features, as shown in Figure 8.27
- hypotonia, often noted at birth
- short small middle phalanx of the fifth finger leading to 'fifth finger clinodactyly'
- congenital heart defects – atrial and ventricular septal defect, common atrioventricular canal and patent ductus arteriosus
- learning difficulties
- increased risk of leukaemia
- premature senescence leading to early cataract formation and Alzheimer disease.

Life expectancy is generally less than 50 years.

HINTS AND TIPS

A genetic disorder is one that is determined by genes, whereas a congenital disorder is one that is present at birth and it may or may not have a genetic basis.

Fig. 8.27 Characteristic features of Down syndrome.

Trisomy 18 – Edwards syndrome

Edwards syndrome arises from trisomy 18 – 47,XX + 18 or 47,XY + 18. The condition has an incidence of 1 in 8000 live births, adjusted for the impact of prenatal screening. Ninety-five per cent die *in utero*, and of those born alive, fewer than 10% survive the first year. The high mortality rate is attributed to the presence of cardiac and renal malformations and a background of apnoea. Characteristic features of ES can be found in Figure 8.28.

Trisomy 13 – Patau syndrome

Patau syndrome (47,XX + 13 or 47,XY + 13) typically arises from non-disjunction. However 20% of cases arise as a result of Robertsonian translocation (see p. 147) as chromosome 13 is acrocentric. PS has an incidence of 1 in 10 000 live births, adjusted for the impact of prenatal screening. Clinical features include:

- dysmorphic features (Fig. 8.29)
- scalp skin defects – aplasia cutis
- incomplete cleavage of the embryonic forebrain – holoprosencephaly
- congenital heart disease.

The median survival age is 2.5 days, with only 1 child in 20 surviving longer than 6 months. Those who survive usually suffer profound mental and physical disabilities.

Sex chromosome disorders

Examples of sex chromosome abnormalities include the following.

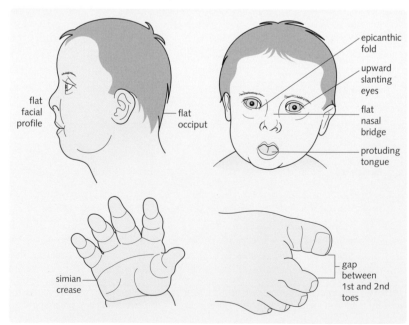

Labels: flat facial profile, flat occiput, epicanthic fold, upward slanting eyes, flat nasal bridge, protuding tongue, simian crease, gap between 1st and 2nd toes

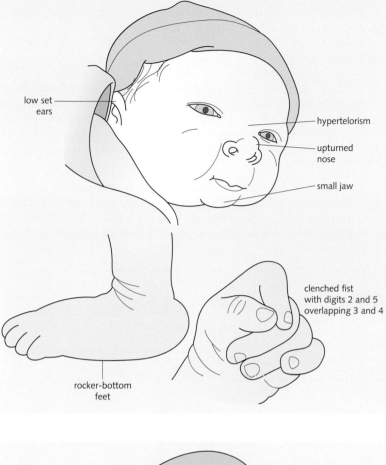

Fig. 8.28 Characteristic features of Edwards syndrome.

low set ears

hypertelorism

upturned nose

small jaw

clenched fist with digits 2 and 5 overlapping 3 and 4

rocker-bottom feet

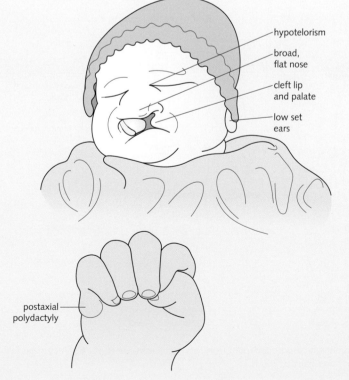

Fig. 8.29 Characteristic features of Patau syndrome.

hypotelorism

broad, flat nose

cleft lip and palate

low set ears

postaxial polydactyly

47,XXY – Klinefelter syndrome

The most common karyotype, 47,XXY, is the most common cause of primary hypogonadism in males and has an incidence of 1 in 1000 live male births. The classic phenotype is shown in Figure 8.30A.

45,X0 – Turner syndrome

Turner syndrome is the only viable monosomy in humans, possibly explained by the fact that the normal situation in a cell is to have just one functional X chromosome, due to lionization. The incidence is 1 in 2500 live female births and the classic phenotype is shown in Figure 8.30B.

47,XXX – Trisomy X syndrome

Trisomy X individuals are usually tall, with a lower than average body mass index. The syndrome presents with a range of phenotypes, with approximately one third exhibiting marked symptoms including infertility, as well as speech and language difficulties. The incidence is approximately 1 in 1500 live female births.

47,XYY

XYY males are often tall, but have normal body proportions. Although often asymptomatic, there may be subtle motor incoordination and behaviour problems. Some present with aggression in childhood. The incidence is 1 in 1000 live male births.

XX male – de la Chapelle syndrome

The XX male phenotype typically arises as a result of unequal crossover between X and Y chromosomes during meiosis. The majority have the Y chromosome gene

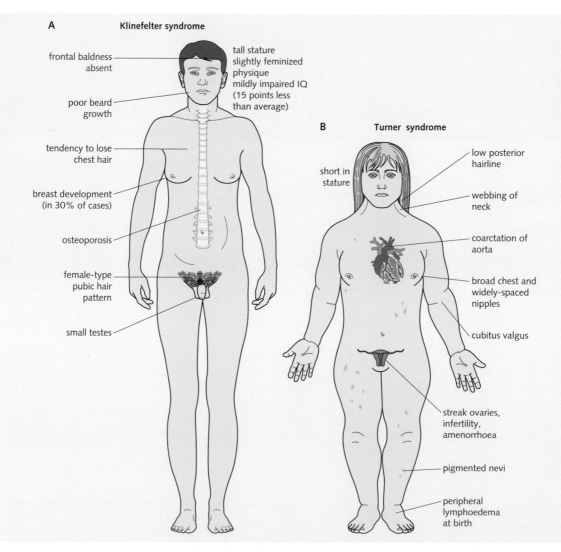

Fig. 8.30 (A) Characteristic features of Klinefelter syndrome (adapted from www.childclinic.net). (B) Characteristic features of Turner syndrome (adapted from Kumar, Abbas and Fausto, 2004).

SRY attached to one of their X chromosomes. They have a similar appearance to males with Klinefelter syndrome, but:

- are sterile
- are usually shorter than average
- have a normal IQ.

XX males usually present at the infertility clinic or when a prenatally predicted female appears to be male. The incidence is 1 in 20 000 live male births.

Mechanisms leading to structural chromosomal disorders

All structural disorders result from chromosomal breakage. Chromosomal damage is increased by some environmental conditions (e.g. mutagenic chemicals, radiation) and by genetic chromosome instability disorders (e.g. ataxia telangiectasia and Fanconi anaemia).

Translocation

Translocation is the exchange of chromosome segments that usually involves dissimilar chromosomes. They may be:

- reciprocal
- Robertsonian.

Translocations may be balanced or unbalanced. In balanced reciprocal translocations there is no gain or loss of genetic material, so the individual is phenotypically normal. However, gametes may result that do not contain a single complete copy of the genome (Fig. 8.31) and so prenatal diagnosis may be offered to known carriers.

A Robertsonian translocation occurs when the long arms of two acrocentric chromosomes (13, 14, 15, 21 and 22) fuse at a centromere, and the two short arms are lost. As these short arms are not known to carry essential genetic material, carriers are phenotypically normal (Fig. 8.32A). Their offspring, however, may inherit a missing or extra long arm of an acrocentric chromosome. Unbalanced Robertsonian translocations occur when two of the same chromosome fuse, causing trisomy or monosomy in the offspring.

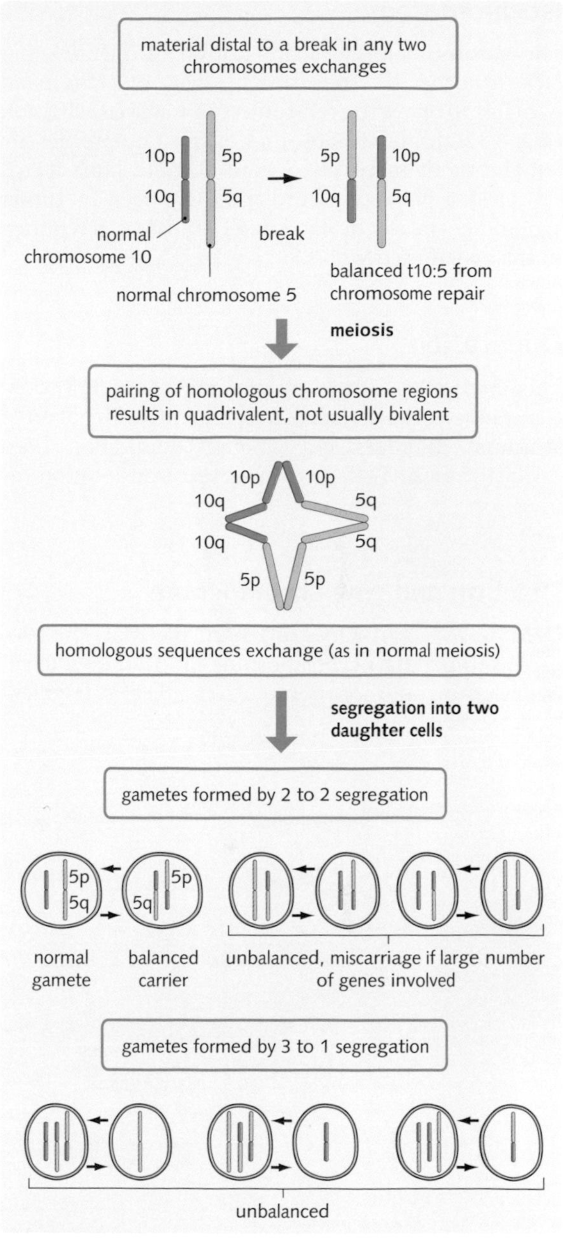

> **HINTS AND TIPS**
>
> Both reciprocal and Robertsonian translocations may be balanced or unbalanced.

Inversion

Inversion arises when two breaks occur in the chromosome and the intervening DNA rotates through 180°. If an inversion includes the centromere it is known as a

Fig. 8.31 Mechanism of reciprocal translocation (highly simplified).

pericentric inversion (Fig. 8.32B). Those that do not arise from breaks in one arm, and are known as paracentric (Fig. 8.32B). Inversions are usually balanced; however, in meiosis homologous chromosomes can line up and pair only if a loop is formed in the region of the inversion. This leads to an increased chance of duplication and deletion, causing unbalanced rearrangements in the offspring.

Isochromosome

This chromosome has a duplication of one arm, but lacks the other. It results from breakage of a chromatid with fusion above the centromere or transverse division (Fig. 8.32C). Isochromosomes of most autosomes are lethal, but those of the X chromosome [46,X,I(Xq)] can yield a phenotype similar to that seen in Turner syndrome, and isochromosome 18 q has been reported in infants afflicted with ES.

Duplication

This implies an extra copy of a chromosome region. Causes include inheritance from parents with balanced structural disorders and *de novo* duplication from unequal crossing-over in meiosis, translocation or inversion.

Deletion and ring chromosome

Deletion results in a loss of genetic material. The telomere is important in chromosome function, so interstitial deletions are more common. If a deleted fragment has no centromere it will be lost during mitosis (Fig. 8.32D).

Ring chromosomes result from breaks near both telomeres of a chromosome, which aberrantly repair to form a ring, with the regions distal to the breaks being lost. If the ring has a centromere, it will be passed through generations in mitosis (Fig. 8.32E).

For a deletion to be considered 'structural', it must be visible by light microscopy (the smallest is approximately 3 Mb). Structural deletions may cause deletion syndromes, in which multiple single gene disorders occur concurrently. Smaller deletions are known as sub-microscopic deletions.

Examples of structural chromosomal disorders

These contiguous gene disorders are diagnosed with the light microscope by high-resolution banding. It is important to determine whether the deletion has arisen from a balanced translocation in the parents (high recurrence risk), or *de novo* (negligible risk). Figure 8.33 describes the features of some structural chromosomal disorders.

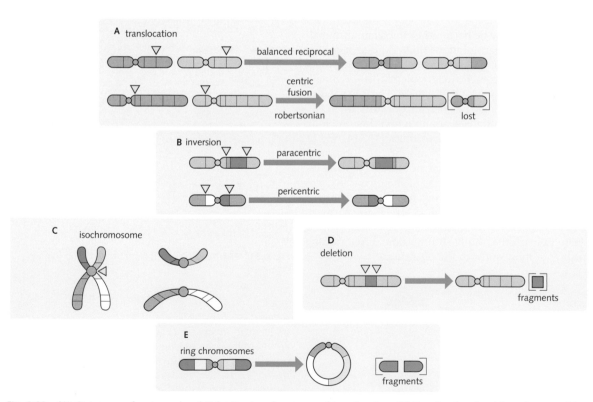

Fig. 8.32 (A) Outcomes of reciprocal and Robertsonian chromosomal translocation. (B) Results of pericentric and paracentric chromosomal inversion. (C) Formation of an isochromosome. (D) Outcome of chromosomal deletions. (E) Formation of a ring chromosome. (Adapted from Kumar, Abbas and Fausto, 2004.)

Fig. 8.33 Features of some structural chromosome abnormalities.

Disorder	Aetiology	Features
Prader–Willi syndrome and Angelman syndrome; incidence 1 in 25 000 live births	Both result from deletion in the same region on 15q11–13; syndromes differ due to genomic imprinting, so depend on which parent the deleted gene is inherited from—Prader–Willi syndrome results from inheritance of the deletion from the father (so only have maternal contribution to the critical area), Angelman syndrome results from inheritance of the deletion from the mother; uniparental disomy can result in these syndromes but does not involve deletions	Prader–Willi syndrome—neonatal hypotonia, initial feeding difficulties, obesity of face, trunk, and limbs (after first year of life), prominent forehead, almond-shaped eyes, triangular upper lips, IQ 20–80, short stature, small hands and feet, hypoplasia of external genitalia, tendency to diabetes mellitus. Angelman (happy puppet syndrome)—hypertonia, ataxic gait, characteristic arm posture, prominent jaw, deep set eyes, happy appearance, laughter, absent speech, learning disability
Wolf–Hirschhorn syndrome	Partial deletion of the short arm of chromosome 4 (4p16.1); male to female ratio is 3:4	'Greek helmet' shaped head, cleft lip and palate, abnormal low-set ears, large beaked nose, hypertelorism (widely spaced eyes), epicanthic folds, microcephaly and learning disability, failure to thrive, heart defects, convulsions, hypospadias
Cri du chat syndrome	Deletion of region on 5p15.2 or the whole short arm of chromosome 5	Round face (in adults the face elongates), cat-like cry (*cri du chat*), hypertelorism, epicanthic folds, strabismus, low-set ears, low birth weight, learning disability (variable degree), appears normal at birth

TREATMENT OF GENETIC DISEASE

Currently, nearly all genetic disorders are incurable, but many are controllable. One of the great hopes of the post-genomic era is that the elusive cures become possible and commonplace.

Conventional strategies

Supportive treatment

Conventional treatment of genetic disorder aims to relieve symptoms and reduce complications. It includes several modalities from pharmacological and surgical treatments to physiotherapy, occupational therapy and speech and language therapy. For example, management of Marfan syndrome requires regular ophthalmology and cardiology review, and may include the prescription of β-blockers and glasses. Surgery and physiotherapy also play an important role.

Surgical correction of gross phenotype

Features of certain syndromes can be corrected surgically for functional and cosmetic reasons, such as the correction of cleft lip and palate.

Environmental modification

The severity and effects of some genetic disorders may be reduced by avoiding key environmental compounds. For example, patients with α_1-antitrypsin deficiency may delay the severity of emphysema by avoiding exposure to tobacco smoke. Patients with glucose-6-phosphate dehydrogenase (G6PD) deficiency are advised to avoid certain drugs (e.g. aspirin).

Substrate limitation

Some genetic disease may be alleviated by limiting exposure to certain substrates.

As phenylketonuria patients do not possess phenylalanine hydroxylase, avoiding dietary exposure to phenylalanine from the first month of life ensures normal intellectual development. Treatment with longterm penicillamine chelates copper and reduces the effects of Wilson disease.

Enzyme and protein replacement

Some enzyme- or protein-deficient conditions are treatable by replacement therapy.

- Type 1 diabetes is therapeutically treated with exogenous recombinant human insulin.

- Haemophilia A is corrected by intravenous infusion of a factor VIII concentrate.
- Type 1 Gaucher disease can be treated with a mannose-6-phosphate modified β-glucosidase, both alleviating symptoms and reducing organomegaly.

However, enzyme replacement therapy is not the magic bullet once envisaged. In the absence of an efficient enzyme delivery system, enzyme replacement therapy has poor results. For example, types 2 and 3 Gaucher disease are not treatable by enzyme replacement therapy due to lack of a suitable CNS delivery mechanism.

Novel and future therapies

Gene therapy

Gene therapy refers to the genetic alteration of an individual's cells in order to correct the underlying genetic abnormality. It could potentially be used to treat many diseases including cystic fibrosis, haemophilia and immune deficiency disorders. It may be achieved by:

- the introduction of the functional gene sequence into target cell DNA
- the introduction of transfected genes expressed from vectors or integrated into the host genome.

Currently, two main methods are utilized for introducing gene sequences:

1. Viral – viruses including lentiviruses, adenoviruses, and herpes viruses have been manipulated to deliver recombinant DNA constructs.
2. Non-viral – non-viral delivery methods include the introduction of naked DNA directly into the cell, and the use of liposome-mediated DNA transfer.

Figure 8.34 compares two delivery vectors that have been trialled for use in cystic fibrosis.

Although, in theory, target cells may be either somatic or germline in nature, the ethics of germline manipulation remains controversial, and current gene-therapy protocols concentrate on somatic intervention.

> **Clinical Note**
>
> ADA-SCID is an AR inherited severe combined immunodeficiency disease (SCID).
>
> One of the earliest somatic gene therapy protocols approved by the US National Institutes of Health was for the use of retroviral vectors to target the ADA gene into patients' *in vitro* stimulated lymphocytes, which were then injected back into the body. As lymphocytes have a limited lifespan, such patients must undergo re-injections every few months, and some have now been in remission for more than 10 years.
>
> More recently, autologous CD34$^+$ bone marrow transplants have been successfully used to induce long-term remission in patients with ADA-SCID, without the need for re-injections.

For gene therapy to become a more powerful treatment method, factors that still need to be surmounted include:

- the short half-life of therapeutic DNA constructs
- the 'host' immune response
- problems with viral vectors – including gene targeting and control issues, and the fear of viral reversion to wild type.

Stem cell therapy

Rather than replacing defective genes and gene products, stem-cell therapy aims to utilize the pluripotency of stem cells to replace defective cells and organ systems. By injecting either embryonic or bone marrow stem cells into the diseased tissue, it is hoped that the stem cells will differentiate into the relevant cell type, replacing the diseased counterparts. Stem-cell therapy may hold the keys to curing, among others, type 1

Fig. 8.34 Advantages and disadvantages of adenovirus and liposome vectors for cystic fibrosis gene therapy.

Vector	Advantages	Disadvantages
Recombinant adenovirus	Targets the respiratory tract epithelium; infects non-replicating cells; in preliminary experiments 40% of respiratory epithelial cells took up the vector	Expression in target cells transient; can create an inflammatory response in recipient; not known whether re-administration is safe (possibility of secondary immune response)
Liposome	Can be delivered directly to the lung (in an aerosol or by direct irrigation or intravenous infusion); re-administration is unlikely to cause an immune response	Low uptake of liposomes: only 5% of epithelial cells transfected in preliminary experiments, which is not enough to have a therapeutic effect

diabetes mellitus and neurological conditions such as Parkinson disease. Stem-cell transplants are already commonplace for some haematological conditions, such as leukaemia.

Clinical Note

Stem cells can be harvested from various locations. Adult stem-cell transplants can be conducted using multipotent bone marrow stem cells from either autologous or allogenic sources. Allogenic stem cells bring with them the necessity for a full HLA match and the risk of graft–versus–host disease (GVHD). Umbilical cord blood stem cells are less differentiated than adult stem cells, less likely to result in GVHD, and do not require a perfect HLA-match.

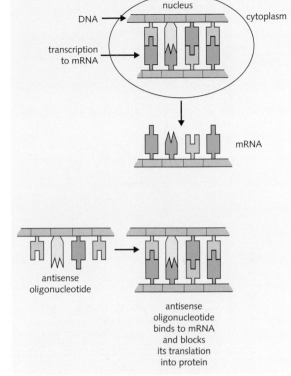

Fig. 8.35 Gene therapy using an antisense technique. Binding of the abnormal mRNA by the antisense molecule prevents its translation into an abnormal protein. Reproduced from Jorde et al., 2006.

Future therapies

A number of technologies have shown therapeutic potential against genetic disease, but are not yet in widespread clinical use. These include the following.

Antisense oligonucleotides

A potential therapeutic target for the treatment of gain-of-function or dominant negative (see p. 129) conditions is the use of anti-sense oligonucleotides (AOs) to bind to mRNA and prevent gene expression (Fig. 8.35). First generation technologies were limited to use at immunologically privileged sites due to degradation by nucleases (e.g. fomivirsen used to treat CMV retinitis). However, second generation AOs consist of nucleic acid analogues which offer more stability.

AOs may also be used to restore function in mutated genes by inducing altered splicing to bypass the disease-causing mutation (exon skipping). This technology has yielded promising results in early trials targeting Duchenne muscular dystrophy (DMD).

Therapeutic siRNA

RNA interference (RNAi) offers the possibility to inhibit gene expression in a sequence-dependent fashion, to selectively 'turn off' the disease gene by using double-stranded RNA (dsRNA). Multicellular organisms utilize an enzyme called 'dicer' to digest viral dsRNA into small fragments, which then become templates to direct the destruction of single-stranded RNA. Synthesized double-stranded RNA molecules (siRNA) that correspond to a disease-causing DNA sequence can therefore be used to destroy specific disease-related mRNA sequences (Fig. 8.36).

Promising therapeutic targets include viral infections, neurodegenerative diseases and some cancers.

Zinc finger nucleases

Zinc finger nucleases (ZFNs) are a type of genetically engineered restriction enzyme designed to recognize long target sequences (which only occur once in the genome) and produce a double-strand cut at that location. Cellular mechanisms repair the double-strand break by copying a donor strand of homologous DNA. This donor strand can be introduced (e.g. on a plasmid containing the wild-type allele from a healthy genome – Fig. 8.37). This technology can also be used to inactivate harmful genes such as the CCR5, which codes for a protein used by HIV to infect lymphocytes.

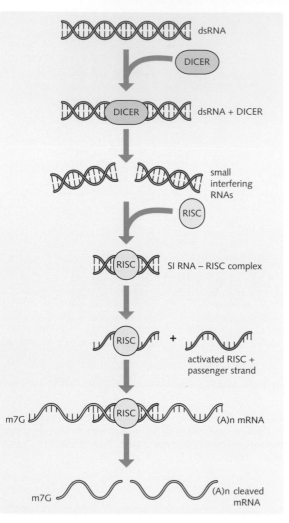

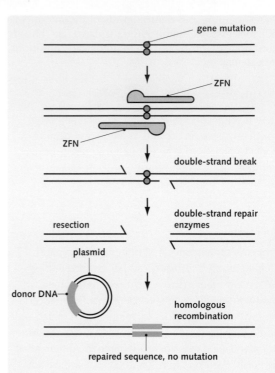

Fig. 8.37 A pair of zinc finger nucleases (ZFNs) bind specifically to 12 bp sequences either side of the genetic mutation, creating a double cut. DNA repair enzymes resect areas either side of the double cut, and repair the damaged area by homologous recombination. A plasmid containing a non-mutated form of the gene acts as a template for repair.

Fig. 8.36 Mechanism of RNA interference. On entering the cell, dsRNAs are processed by dicer (a RNAse III enzyme), producing siRNA. The siRNA is incorporated into RISC, which cleaves the siRNA and discards the passenger (sense) strand. The anti-sense strand of the siRNA targets RISC to its homologous mRNA sequence, resulting in endonucleolytic cleavage of the targeted mRNA. siRNA, short interfering RNA; RISC, RNA-inducing silencing complex.

Principles of medical genetics ⑨

Objectives

By the end of this chapter you should be able to:
- Define the Hardy–Weinberg equilibrium and list the factors that may disturb it.
- List the Wilson and Junger screening criteria.
- Understand the purpose of carrier detection and give examples of its use.
- Appreciate the importance of maternal serum screening and the use of multiples of the median calculations.
- Understand the risk of a child being affected by an autosomal recessive condition if both parents are carriers, and the probability of a child inheriting an autosomal dominant condition if one parent is affected.
- Understand how Bayes' theorem is used in genetic counselling.
- Appreciate the importance of a good family history in the context of the genetic consultation.
- Consider some of the ethical implications in medical and clinical genetics.

POPULATION GENETICS AND SCREENING

It is estimated that between 2 and 5% of new-borns have congenital malformations or genetic disorders. Many common diseases also have a genetic component. The role of the clinical geneticist in the management of individuals and families affected by a genetic disease, includes:

- establishing an accurate diagnosis
- providing information about prognosis
- calculating the risks of developing or transmitting the disorder
- exploring ways in which the development or transmission of the disorder may be modified.

Without established therapies, the emphasis is placed on identifying individuals who are at risk of having an affected child so that they can make informed reproductive choices. Such individuals may be identified:

- in families where a genetic disorder has arisen
- in at-risk groups (e.g. Tay–Sachs disease in Ashkenazi Jews).

Population genetics

Population genetics is the study of the genetic composition of populations. Allele frequencies and the frequency of disease-causing mutations vary considerably between populations. Cystic fibrosis is common in Europeans of Celtic and Northern European descent

(allele frequency 1/40–1/50), but rare in Finnish, Asian and African populations; while Tay–Sachs disease has a high allele frequency in Ashkenazi Jews (1/60), 100-fold more common than in other populations.

In order for clinical geneticists to accurately assess risk, they must know how common the disease-causing mutation is in the relevant population, particularly for recessive disorders.

The Hardy–Weinberg principle enables the frequency of a disease-causing allele in a population to be calculated from the disease incidence (provided that the population is in Hardy–Weinberg equilibrium).

Hardy–Weinberg law and equilibrium

The Hardy–Weinberg law states that allele frequencies within a population tend to remain constant from one generation to the next. It assumes that the organism under investigation is diploid, is sexually reproducing and has discrete generations. By extension, the Hardy–Weinberg equilibrium states that allele frequencies remain constant from one generation to the next, providing there is an absence of selection. Thus, conditions required to satisfy the equation are:

- the population is large, to minimize genetic drift
- mating is random
- the mutation rate remains constant
- alleles are not selected for (i.e. confer no survival or reproductive advantage)
- there is no migration into or out of the population.

153

Fig. 9.1 The Hardy–Weinberg equation. If two heterozygotes mate, Aa × Aa. The distribution of AA, Aa, and aa genotypes in the population correspond to p^2, 2 pq, and q^2, respectively. If A and a are the only alternative alleles for the same gene locus, then, $p^2 + 2\,pq + q^2 = 1$. a, recessive allele; A, dominant allele; p, frequency of A; q, frequency of a.

The Hardy–Weinberg equation is derived by considering a population carrying an autosomal gene with two alleles A and a. The frequency of the dominant allele A in gametes is represented by p, and the frequency of the recessive allele a is represented by q (Fig. 9.1). Since there are only two alleles, $p + q = 1$ and $q = 1 - p$.

> **HINTS AND TIPS**
>
> Information on the diagnosis, management and counselling of specific genetic disorders can be found on http://www.geneclinics.org/.

In the combination of alleles that forms the next generation:

- the chance that both male and female gametes will carry the A allele is $p \times p = p^2$
- the chance that the gametes will produce a heterozygote is $(p \times q) + (q \times p) = 2\,pq$
- the chance that both male and female gametes will carry the a allele is $q \times q = q^2$.

These are the only possibilities, therefore:

$$p^2 + 2\,pq + q^2 = 1$$

This equation can be used to calculate allele frequency in a population if disease occurrence is known, provided that the population is in Hardy–Weinberg equilibrium.

For autosomal recessive (AR) conditions disease incidence $= q^2$; gene frequency $= q$; heterozygote frequency $= 2\,pq$. For example, an AR disorder occurs with a frequency of 1 in 1600 live born births:

- incidence, $q^2 = 1/1600$
- gene frequency (a), $q = 1/40$; dominant allele A has gene frequency $p = 39/40$
- heterozygote frequency, 2 pq is $2 \times 39/40 \times 1/40$, which is approximately 1/20.

For autosomal dominant (AD) conditions:

- nearly all affected are heterozygotes, so q^2 is approximately 0
- if the condition is rare, p^2 is approximately 1
- disease gene frequency (A) is approximately 2 pq, which is approximately 2 q.

Factors which disturb Hardy–Weinberg equilibrium

A population is not in Hardy–Weinberg equilibrium if the genotype frequencies do not arise in the proportions predicted by the Hardy–Weinberg law (Fig. 9.2A, 9.2B). This may arise as a result of a number of mechanisms, discussed below.

Non-random mating

Random mating refers to the selection of a partner regardless of genotype. The tendency to select partners who share certain characteristics (e.g. height) is called assortative mating, which is non-random. Non-random mating may also result from consanguinity.

Mutation

If a locus has a high mutation rate, theoretically there will be an increase in the number of mutant alleles in the population. In practice, this does not occur for populations in Hardy–Weinberg equilibrium, because the mutation is selected against.

Selection

The reduced reproductive fitness of affected individuals acts as a negative selection, which leads to a gradual reduction in the frequency of the mutated gene. In practice, this is balanced by the mutation rate for populations in Hardy–Weinberg equilibrium.

In some cases, selection of a mutation may increase fitness and, for some autosomal recessive disorders, heterozygotes show an increase in fitness relative to unaffected homozygotes (the 'heterozygous advantage'). For example, carriers of sickle-cell anaemia are comparatively resistant to malaria.

Genetic drift

In small populations, one allele may be transmitted to a high proportion of offspring by chance, resulting in marked changes in allele frequency between the two generations and a disturbance in Hardy–Weinberg equilibrium. This phenomenon may contribute to the 'founder effect' (see below).

Migration

The introduction into the population of new alleles as a result of migration and subsequent intermarriage will result in a change in the relevant allele frequencies, and a disturbance to Hardy–Weinberg equilibrium.

Fig. 9.2A Determining whether a population is in Hardy–Weinberg equilibrium given the genotype frequencies. The genotype frequencies are used to determine the allele frequencies.

	Genotype			
	TT	Tt	tt	Total
No. of individuals	40	40	20	100
No. of T allcles	80	40	0	120
Nu. of t alleles	0	40	40	80

Total number of alleles = 120 + 80 = 200
Frequency of T = p = 120/200 = 0.6
Frequency of t = q = 80/200 = 0.4

Fig. 9.2B Chi squared (χ2) tests are used to determine whether the population differs significantly from one in Hardy–Weinberg equilibrium. The observed (O) genotype frequencies are those seen in the population. The expected (E) genotype frequencies are those predicted if the population is in Hardy–Weinberg equilibrium, using the allele frequencies calculated in (A). From χ2 tables (with one degree of freedom and 95% confidence intervals), a population does not differ significantly from one in Hardy–Weinberg equilibrium if χ2 is less than 3.84.

	Genotype		
	TT	Tt	tt
Observed (O)	40	40	20
Expected (E)	$p^2 \times 100 = 36$	$2pq \times 100 = 48$	$q^2 \times 100 = 16$
O – E	4	–8	4
$(O - E)^2/E$	16/36 = 0.44	64/48 = 1.33	16/16 = 1

$\chi^2 = \Sigma(O–E)^2/E = 0.44 + 1.33 + 1 = 2.77$

Founder effect

Founder effects arise in populations established from a small group of individuals ('founders'). If, by chance, one founder has a certain disease gene, this will remain over-represented in successive generations, particularly in isolated populations that breed amongst themselves. For example, several rare autosomal recessive disorders occur at a relatively high frequency amongst the Old Order Amish (an isolated group that tend to intermarry).

Population screening and carrier detection

Screening of all members of a population regardless of their family history is used to identify carriers of recessive traits and to detect pre-symptomatic individuals.

The aim of carrier detection is to identify asymptomatic heterozygotes for autosomal recessive (AR) traits, although it is sometimes used to detect carriers of autosomal dominant (AD) disorders that have limited penetrance or late onset (e.g. Huntington disease). This allows counselling and prenatal tests to be offered as appropriate. Carrier detection tends to be confined to small ethnic populations in which there is an anomalously high incidence of a particular disease.

Prenatal diagnosis concerns the use of tests in pregnancy to determine whether an unborn child is affected with a particular disorder. Figure 9.3 describes some of the tests available.

Criteria for a screening programme

Population wide screening programmes must usually meet the criteria laid down by the UK National Screening Committee (NSC). They are an extension of the Wilson–Jungner criteria (WHO 1968).

- The condition being screened for should be an important problem.
- There should be a latent or early symptomatic stage.
- The natural history of the condition should be understood.
- There should be a definitive test or examination for the condition.
- The test or examination should be acceptable to the population.
- Case finding should be a continuing process rather than a 'one-off' project.
- There should be an effective treatment for the disease.
- Facilities for diagnosis and treatment should be available.

Fig. 9.3 Some of the tests available in prenatal diagnosis. (AFP, α-fetoprotein; HCG, human chorionic gonadotrophin; uE3, uncongugated oestradiol; DIA, dimeric inhibin-A; NTDs, neural tube defects; TORCH, toxoplasmosis, other agents, rubella, cytomegalovirus, herpes simplex.) Associated risk refers to the risk that the test may damage the foetus. *The constituents of the quadruple test vary across the UK.

Test	Gestation	Procedure	Abnormalities detected	Risk of procedure
Amniocentesis	16–18 weeks (routinely) 14–32 weeks is possible	Liquor is removed via a long needle inserted transabdominally. Cells are cultured for 2–3 weeks	Foetal sexing, karyotyping, and enzyme assay	Miscarriage rate estimated at 1% above the normal rate at 16 weeks
Chorionic villus sampling	8–10 weeks	Biopsy usually taken transabdominally or transvaginally. Cells are cultured for 2–3 weeks	Foetal sexing, foetal karyotyping, biochemical studies, DNA analysis (cell culture not necessary)	Miscarriage rate estimated at 2–3% above the average at 10 weeks; rhesus isoimmunization if mother is rhesus negative
Cordocentesis	>18 weeks	Foetal blood sample obtained by inserting fine needle into foetal umbilical cord	Suspected foetal infection, Unexplained hydrops, blood disorders.	Procedure-related loss approximately 1%; risk of rhesus isoimmunization
Ultrasound	Routine scan at 16–18 weeks in all pregnancies	Visualization of foetus by transabdominal ultrasound probe	Over 280 congenital malformations	Non-invasive test with low associated risk
Maternal serum screening: infectious screen (TORCH screen), quadruple test* (serum AFP, HCG, uE3 and DIA)	13–26 weeks	Sample of maternal blood collected	Detection of infections that may cause congenital malformations (see Chapter 8); quadruple test estimates relative risk of NTDs and Down syndrome based on serum levels of HCG, AFP, uE3 and DIA	Very low

- There should be an agreed policy on whom to treat as patients.
- The cost of screening should be balanced in relation to overall health expenditure.

Methods used for carrier detections and pre-symptomatic diagnosis

Direct mutation detection
If the mutation(s) that cause a specific disease is known, carriers or pre-symptomatic individuals can be identified by direct mutation analysis using techniques, such as PCR (see Ch. 7) or microarray analysis (see below). Often, there are many mutations that can cause the same disease (allelic heterogenicity). For example, Delta F508 is the most common mutation causing cystic fibrosis (CF), but over a thousand have been identified.

Linkage to a polymorphic marker
Genetic linkage is the tendency for alleles close together on the same chromosome to be transmitted together through meiosis (see Ch. 6). It allows disease genes for which the causal mutation is unknown to be followed through generations, by using a linked polymorphic marker. The process has fairly high error rates however, and can only be reliably used in large families.

Biochemical tests
Over 100 inborn errors of metabolism can be detected by enzyme assays using cultured amniocytes or chorionic villus samples, although most are not used routinely. New-born babies can be screened for errors of metabolism by simple biochemical tests on a blood sample.

Clinical Note

As part of the NHS new-born screening programme, babies born in the UK are routinely screened for a number of conditions. A sample of blood is taken from a heel prick and placed on a Guthrie card, which is analysed in the laboratory. The conditions screened for are phenylketonuria (PKU), congenital hypothyroidism (CHT) and cystic fibrosis (CF). Babies in England, Scotland and Northern Ireland are also screened for sickle-cell disease (SCD) and medium-chain acyl-CoA dehydrogenase deficiency (MCADD), whereas male babies in Wales are screened for Duchenne muscular dystrophy (DMD).

DNA microarrays

DNA microarrays can screen for thousands of genetic conditions simultaneously (see p. 112). These 'gene chips' are likely to play an important role in carrier detection and presymptomatic diagnosis in future, although are not yet widely used.

Screening in 'at risk' populations

Haemoglobinopathies

As part of the NHS sickle-cell disease and thalassaemia screening programme, all new-born babies in England, Scotland and Northern Ireland are now screened for sickle-cell disease. Traditionally however, screening was aimed towards at risk populations. Adult carrier detection is by the Sickledex test.

Traditionally, screening for thalassaemia has been targeted to those of Mediterranean or South East Asian descent, and achieved by measurement of red cell indices and electrophoresis for abnormal haemoglobins.

Tay–Sachs disease

Carriers are detected by hexosaminidase A (HEXA) enzymatic activity in plasma or white cells. Alternatively, direct mutation detection may be offered. These tests are targeted at the Ashkenazi Jewish population.

Cystic fibrosis

Cystic fibrosis (CF) is the most common serious AR condition to affect Caucasians (carrier frequency approximately 1 in 22 in the UK). Adult carrier detection is usually limited to affected families, but three main screening modalities are employed.

- Antenatal screening – via amniocentesis or chorionic villus sampling is offered to parents known to be carriers.
- New-born testing – offered routinely as part of the 'blood spot' protocol, followed by confirmation by mutation analysis.
- Mutation detection – approximately 85% of CF carriers have a 'common' identifiable mutation.

Antenatal screening tests

These are usually simple tests, often of maternal serum, that can be used to assess a woman's relative risk of having an affected child.

At 13–26 weeks' gestation, women can be tested for maternal serum α-fetoprotein (MSAFP), human gonadotrophin (HCG), unconjugated oestradiol (uE3) and dimeric inhibin-A (DIA) (Fig. 9.4). Measurements are corrected for maternal and gestational age, and a multiple of the median (MOM) value calculated to ascertain the relative risk of a number of foetal conditions

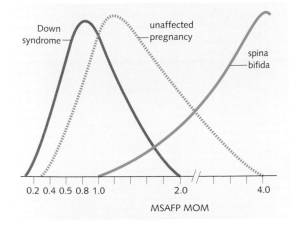

Fig. 9.4 Screening for trisomy 21. Maternal serum α-fetoprotein (MSAFP) levels are used to generate likelihood ratios, derived from overlapping distribution of affected and unaffected pregnancies. Likelihood ratio and maternal age are used together to generate a combined risk. If there is a high combined risk, refer for diagnostic testing. MOM, multiples of the median.

including neural tube defects (spina bifida) and trisomy 21 (Down syndrome) (Fig. 9.5).

- MSAFP is usually low in trisomies 18 and 21, and normal in trisomy 13.
- MSAFP is usually high in neural tube defects.
- uE3 is usually low in trisomies 13, 18 and 21.
- HCG is usually high in trisomy 21 and low in trisomies 18 and 13.
- DIA is usually high in trisomy 21.

'Cascade' screening

Cascade screening is the process of systematically approaching relatives of a patient carrying a disease allele (the 'index case') in order to identify other carriers or those at risk. Once an individual has been diagnosed, family members at high risk are offered testing. If more people within the family are identified with the gene, the results may be used to identify further family members at risk, who are then offered the test – and so on.

RISK ASSESSMENT AND GENETIC COUNSELLING

For families in which a disease is showing a recognizable pattern of Mendelian inheritance, the risk to other family members of developing or transmitting the disorder can be calculated by simple probability theory.

Fig. 9.5 Detection rates using different Down syndrome screening strategies. AFP, α-fetoprotein; HCG, human chorionic gonadotrophin; uE3, uncongugated oestradiol; DIA, dimeric inhibin-A; NT, nuchal translucency.

Screening modalities	% of all pregnancies tested	% of Down syndrome cases detected
Age alone		
40 years and over	1.5	15
35 years and over	7	35
Age + AFP	5	34
Age + AFP, uE3 + HCG	5	61
Age + AFP, uE3, HCG + DIA	5	75
NT alone	5	61
NT + age	5	69
HCG, AFP + age	5	73
NT + AFP, HCG + age	5	86

Probability theory as applied to genetics

The probability of an event occurring can be expressed as a fraction:

p = number of ways events can happen/
total number of possibilities

Probability is a useful way of quantifying risk.

Laws of addition and multiplication of probability

The law of addition

If two events could not happen at the same time, they are said to be mutually exclusive. The probability that either event will occur is equal to the sum of their probabilities.

$$p = p1 + p2$$

The law of addition is sometimes called the 'or' law because it is the probability that one event or another event will occur.

The law of multiplication

An independent event is one that has no effect on subsequent events. The outcome of the first event has no effect on subsequent events. The probability that all the events will occur is equal to the product of the individual probabilities.

$$p = p1 \times p2$$

The law of multiplication is sometimes called the 'and' law because it is the probability that one event and another event will occur.

HINTS AND TIPS

To determine the probability of two independent events both occurring, you should multiply the probabilities of the individual events together.

Calculating risks from pedigree information

Autosomal recessive conditions

AR disorders are only manifested if two mutant copies of the gene are inherited (see Fig. 8.8). Therefore, for a child to have a recessive disorder both parents must be carriers.

HINTS AND TIPS

Read questions on autosomal recessive (AR) disease carefully. For example, if you are asked to define the probability that a healthy child sired by two carrier parents is a carrier for cystic fibrosis, the answer is {2/3}, because you know the child does not have the condition. However, if you are asked to calculate the probability that an unborn child will be a carrier the answer is {1/2} because they might also be affected.

Autosomal dominant conditions

AD conditions are manifested if one mutant gene is inherited (see Fig. 8.5). Therefore, for a child to have a dominant disorder, either the mother OR the father must have the disease gene, or the disease could arise *de novo*.

X-linked recessive conditions

The situation is more complicated with X-linked recessive disorders because the situation varies according to the sex of the child (see Fig. 8.11).

Bayes' theorem

Additional information may be used to modify the risk calculated from the pedigree data in order to obtain a more accurate value. For example, given that Huntington disease generally manifests in middle age, a suspected carrier who has not yet developed the disease at age 30 may still have the mutation. However, if they have not developed the disease by age 60, it is less likely that they carry the mutation. Bayesian analysis takes such considerations into account by determining the relative probabilities of two alternative outcomes.

- The 'prior' probability is based on classic Mendelian inheritance.
- The 'conditional' probability is based on observations that modify the prior probability, such as unaffected offspring, results of screening tests, or age.

A 'joint' probability is then calculated as the product of the prior and the conditional probability. A final 'relative' probability is the proportional risk of one alternative with respect to the other (Figs 9.6, 9.7).

X-linked recessive disorder—haemophilia A

A

B relative risk calculation using a Bayes' table

II$_2$	P (carrier)	P (non-carrier)	
prior	$1/2$	$1/2$	(Mendelian calculation)
conditional	$1/2 \times 1/2 \times 1/2 = 1/8$	1	(based on existing information—three unaffected sons)
joint	$1/2 \times 1/8 = 1/16$	$1/2$	(product of prior and conditional)
relative	$1/16 : 1/2 = 1 : 8 = 1/9$	$1/2 : 1/16 = 8 : 1 = 8/9$	(relative risk, based on joint probabilities)

Fig. 9.6 Estimation of carrier relative risk using Bayes' theorem. (A) Mother (I$_2$) is an obligate carrier as she has an affected brother and son. Daughter (II$_2$) has already had three unaffected sons. (B) The relative risk can be displayed on a Bayes' table.

---

isolated X-linked recessive case — Duchenne muscular dystrophy (DMD)

A

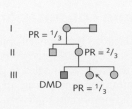

B relative risk calculation using a Bayes' table

III$_2$	P (carrier)	P (non-carrier)
prior	$1/3$	$2/3$
conditional	$1/3$	1 *
joint	$1/3 \times 1/3 = 1/9$	$2/3 \times 1 = 2/3$
relative	$1/9 : 2/3 = 1 : 7$	$2/3 : 1/9 = 7 : 1$

* all non-carriers have normal CK test

Fig. 9.7 Estimation of carrier proportional risk using Bayes' theorem. (A) Isolated case of DMD in III$_1$; assess carrier risk in his sister, III$_2$. One-third of cases arise from new mutations – neither mother nor sister would undergo mutation. If mutation was inherited from the mother, the carrier risk for the sister would be half. If the mutation was from the grandmother, the mother would be an obligate carrier (carrier risk 1) and the daughter's carrier risk would be half. Based on this, prior risk for the grandmother and daughter are one-third, and the mother is two-thirds. The daughter has a normal creatine kinase (CK) test. In general, two-thirds of carriers have raised CK and one-third have normal CK. (B) The proportional risk can be displayed on a Bayes' table.

Aspects of genetic counselling

Establishing a diagnosis

Genetic counselling is the provision of information to affected individuals or family members at risk of a genetic disorder. The consultands are informed of:

- the consequences of the disorder
- the probability of developing or transmitting it
- ways in which it may be prevented or ameliorated.

An accurate diagnosis is essential so that the correct advice can be given. A medical history of all affected family members is needed and a pedigree constructed. Miscarriages, unexplained learning difficulties, congenital malformations and parental consanguinity should be specifically enquired about.

Investigations may involve chromosomal or DNA analysis, or specific biochemical tests.

Presenting the risks in context

Once the diagnosis and mode of inheritance have been established, carrier risk and recurrence risk can be estimated, based on Mendelian rules and Bayes' theorem.

Discussing options, communication and support

The counsellor must aim to be non-judgemental and non-directive towards the consultands when discussing their future options. At best, the counsellor can give the consultand reassurance that the recurrence risk is no greater than the population risk. If the recurrence risk is high, the counsellor should explore the emotional, physical and financial implications, and offer information regarding:

- further pregnancies – contraception advice may be appropriate
- prenatal diagnosis – consider selective termination
- artificial insemination of donor sperm (AID) – if the male has an AD condition or both partners are carriers for an AR condition
- *in vitro* fertilization (IVF) with pre-implantation genetic diagnosis
- egg donation – in the case of AR, X-linked, or maternal mitochondrial disorders
- accepting the risk and coping if an affected child is born
- other options, including adoption.

Follow-up sessions should always be offered to give the consultands time to absorb the information and the opportunity to ask further questions.

Clinical Note

Remember 'ICE': consultands Ideas, Concerns and Expectations should always be explored. People need time to make rational, well thought-out decisions. It is the counsellor's job to ensure these decisions are well informed.

Ethical considerations in genetic counselling

Consanguinity and incest

Incest is the mating of first-degree relatives. In the UK, double first cousins are the closest relatives allowed to marry (i.e. the two sets of parents are both full siblings). The risk of disease or serious congenital malformation in a child born to first cousins is 1 in 20, and even higher in a highly inbred family. A detailed anomalies scan and careful monitoring are therefore indicated.

Disputed paternity

Paternity testing uses PCR analysis (see Ch. 7) to determine paternity with an extremely low error rate. However, the social connotations for the families involved may be profound, and sensitivity and discretion are vital.

Confidentiality and conflicts of interest

Patient confidentiality is vital. However, this can sometimes be a problem if a family has a disease with a very clear pattern of inheritance and only some members of the family want to know their risks (see p. 130).

HISTORY AND EXAMINATION

History and consultation

An accurate diagnosis is essential if appropriate advice is to be given. Some superficially similar disorders may have totally different aetiologies (e.g. dwarfism) and incorrect assumptions may lead to misleading risk calculations. It is important to determine what the consultands know about genetics, and provide information in a way they can understand.

Outline

The outline of a genetic history is in many ways similar as to that for any medical condition. However, certain areas are crucial, and should be explored in detail.

If the patient is a child, the history will be mostly obtained from the parents. However, collateral history from teachers or other carers may prove useful.

As with any history, the presenting complaint is vital. It is particularly important to explore the time-course of the complaint, and any associated symptoms.

Particular attention should be given to family history, and a family tree should be constructed to show how the condition has been passed on (see Fig. 8.4). It may reveal the mode of inheritance, penetrance and possibly where the disease originated from. In this way, realistic risk calculations can be made.

Social, drug and past medical history are all important as these often give a clue as to the presence or absence of a genetic syndrome. Patients may not have connected problems they had in the past, or in their social lives, with the complaint they are now presenting with. Similarly, a comprehensive review of systems is vital.

Specific lines of inquiry

A few questions may be difficult to ask but are particularly important. Enquire specifically about:

- infant deaths, stillbirths and abortions as this may alter recurrence risks (e.g. spina bifida is associated with an increased risk of neural tube defect in subsequent children)
- consanguinity
- non-paternity (discreetly!), as this may explain unexpected disease incidences
- ethnicity and country of descent
- obstetric history (e.g. maternal health, teratogen exposure and viral infections).

Examination

Genetic conditions can cause a huge variety of abnormalities, and so a thorough and systematic systems-based approach should be adopted. Although some conditions may be readily identifiable, in general the findings at examination should be used, along with the history, to guide further investigation in order to make the diagnosis. Some common presentations of genetic disease are given in Figure 9.8.

ETHICAL ISSUES IN MEDICAL GENETICS

The discipline of genetics attracts ethical debate perhaps unlike any other specialty. Advents in technology now allows us to cross the border past what is generally taken as morally and ethically acceptable. (A full and comprehensive introduction to ethics, ethical theories and practice can be found in *Crash Course: Ethics and Human Sciences*.)

Medical ethics and its principles

In medical ethics there are four basic principles:

1. The principle of beneficence – the principle of seeking to do good.
2. The principle of non-maleficence – the principle of seeking, overall, to do no harm.
3. The principle of justice – incorporating fairness in the context of the resources available, equity of access and opportunity.
4. Respect for patients' autonomy – incorporating respect for the individual and their decisions, informed consent and confidentiality.

Genethics

Ethical dilemmas encountered in medical genetics highlight the importance of the law and public scrutiny, as well as genetic counselling (see p. 157). They include:

- presymptomatic and susceptibility testing
- testing of children and adolescents
- preimplantation diagnosis
- prenatal diagnosis and therapeutic termination
- genetic testing for non-disease traits
- data protection and ownership of genetic data.

Presymptomatic and susceptibility testing

Those individuals known to be at risk of genetic disorder, especially late-onset disorders, may be offered susceptibility testing. The usefulness of such tests is affected by the clinical course of the disease and availability of treatment, and the individual's response to the result.

For conditions that do not display full penetrance, a serious limitation of susceptibility testing is interpretation. A positive result only confers that there is a risk, and does not mean that the person will necessarily develop disease.

Familial cancers

In the case of familial cancers, a positive result for mutation in causative genes alters the frequency of screening, allowing earlier detection and improving prognosis. It also allows an individual the chance to consider therapeutic surgical options, such as mastectomy in the case of familial breast cancer.

Late onset diseases

Susceptibility testing is available for a number of conditions which present late in life including polycystic kidneys, familial Alzheimer, Parkinson and Huntington disease. There are currently few effective treatments to halt these from progressing, however definitive

Fig. 9.8 Common presentations and associations of genetic disease.

Area/system	Example	Association
Development	Social skills	Gregarious personality in Williams syndrome Inappropriate laughter and absent speech in Angelman syndrome
	Failure to thrive	Pancreatic insufficiency of cystic fibrosis. Metabolic disorders (e.g. mucopolysaccharidoses)
Oral and facial deformities	Hypertelorism	Basal cell nevus syndrome, DiGeorge syndrome and Loeys–Dietz syndrome
	Cleft lip and palate	Chondrodysplasia punctata, trisomy 13 and Pierre Robin syndrome
	Craniosynostosis	Crouzon syndrome
	Low set ears	Patau syndrome, Edwards syndrome, Turner syndrome
Skin	Ichthyoses	Chondrodysplasia punctate, Conradi syndrome
	Blistering	Epidermolysis bullosa
Bone and connective tissue	Polydactyly	Patau syndrome; Ellis–van Creveld syndrome
	Brachydactyly	Turner syndrome
	Rockerbottom feet	Edwards syndrome
Congenital heart disease	Pulmonary stenosis and atrial septal defects	Noonan syndrome
	Aortic stenosis and pulmonary stenosis	Williams syndrome
	Coarctation of the aorta	Turner syndrome
	Truncus arteriosus and pulmonary atresia	DiGeorge syndrome
Gastrointestinal disorders	Exomphalos and gastroschisis-exomphalos	Beckwith syndrome
	Meconium ileus	Cystic fibrosis
	Duodenal atresia	Down syndrome
	Imperforate anus	VATER syndrome
Liver disease	Cirrhosis	Wilson disease, haemochromatosis
	Cirrhosis and emphysema	a_1-antitrypsin deficiency
	Isolated jaundice	Gilbert's syndrome
Genitourinary tract disorders	Polycystic kidney disease	Patau syndrome, Edwards syndrome, tuberous sclerosis, Meckel syndrome
	Renal tumours	Wilms' tumour, von Hippel–Lindau syndrome
	Hypogonadism	Klinefelter syndrome, Tuner syndrome and Prader-Willi syndrome
Blood disorders	Pancytopenia	Fanconi anaemia
	Disordered coagulation	Haemophilia A and B, von Willebrand disease
	Immunodeficiency	X-linked Bruton's agammaglobulinaemia, servere combined immunodeficiency

diagnosis may allow an individual to make certain life choices, for instance taking the decision not to have children, or to investigate preimplantation diagnostics before starting a family. Others prefer not to know.

Clinical Note

Mr X Snr has Huntingdon disease. His granddaughter is thinking of starting a family and wishes to take the test to find out if she is affected before making a decision. Her father, Mr X Snr's son, does not want to be tested or to know his disease status.

This scenario raises the issue of testing 'by proxy'. If Ms X takes the test, and is positive, it would imply that her father is also positive and is likely to develop the disease. Non-disclosure of such results is difficult within immediate families and this may create severe family tension.

Testing of children and adolescents

Issues surrounding the testing of children are, in part, the same as those for adults, namely the usefulness of the test in guiding treatment versus the identification of autosomal dominant disease for which there is no known treatment. While the former has very clear benefit, the later constitutes a definite infringement of a child's future autonomy. For example, in the case of children with a mutation in the adenomatous polyposis coli (*APC*) gene, screening for colon cancer by flexible sigmoidoscopy or colonoscopy is indicated from an early age and is potentially lifesaving. In the instance of fatal adult-onset autosomal conditions, diagnosis while a child may have a great psychological impact and lead to stigmatization.

Pre-implantation diagnosis

Pre-implantation genetic diagnosis (PGD) is used to diagnose a severe genetic or chromosomal condition in an *in vitro* fertilized embryo. Its use has raised questions about genetic selection, the social power of genetic information and the devaluation of human life. The UK Human Fertilisation and Embryology Authority govern which conditions may be identified by means of PGD (Fig. 9.9).

PGD has also has the power to identify an embryo that can serve as a tissue match for a sick child, the so-called 'saviour sibling'. There is no preset list of conditions for which this is permitted, and each family has to apply individually. This process includes a review of psychological and emotional implications for each child and their families.

HINTS AND TIPS

The Human Fertilisation and Embryology Authority (HFEA) has a number of important roles:
- to provide information
- to licence and regulate embryo research
- to regulate the use of fertility and embryo-related technologies (including PGD).

There are currently government plans to disband HFEA and devolve these roles to other organizations including the Quality Care Commission and a new medical research agency – watch this space.

Prenatal diagnosis and therapeutic termination of pregnancy

Prenatal diagnosis of structural and genetic abnormalities may influence:

- the management of the pregnancy, including planning for possible complications at birth
- the strategy for dealing with problems that may occur in the new-born
- the decision whether or not to continue the pregnancy
- the decision whether or not to have more children in the future.

Ethical considerations of prenatal diagnosis include:

- who should be screened and what is the screening threshold that warrants diagnostic testing?
- informed consent – can someone truly understand the implications of every condition that may be detected before consenting?
- the inherent risk of miscarriage associated with some diagnostic procedures
- the accuracy of the tests
- when to consider termination, what constitutes a 'serious' defect?
- the psychological impact of making a 'life or death' decision with regard to your unborn child
- the eradication of disease from the population by 'screening eugenics' and the devaluation of human life.

HINTS AND TIPS

In the UK termination of pregnancy is permitted up to 24 weeks' gestation, under statutory grounds C and D of the Abortion Law Act, 1967. However, this may be extended beyond 24 weeks if the foetus has a lethal condition, if there is a substantial risk that the resulting child would be born with serious handicap or if there is

Fig. 9.9 Some of the conditions for which the use of PGD has been licensed by HFEA. A total of more than 100 conditions has been licensed, and many more are being considered. Prevalence quoted is approximate per 100 000 in Europe. Mode, mode of inheritance; ARNSSD, Autosomal recessive non-syndromic sensorineural deafness.

Mode	Condition	Prevalence (per 100 000)
Autosomal	Charcot–Marie–Tooth disease	33
Dominant	Huntington disease	6
	Neurofibromatosis I and II	26
	Osteogenesis imperfecta	7
	Marfan syndrome	20
	Hereditary breast - ovarian cancer syndrome (BRCA I + II)	25
Autosomal	Phenylketonuria	4
Recessive	β-thalasaemia	0.5
	Cystic fibrosis	12
	Tay–Sachs disease	0.3
	Sickle-cell disease	11
	Spinal muscular atrophy type I	1
X–linked	Adrenoleukodystrophy	5
	Duchenne muscular dystrophy	3
	Fabry disease	2
	Fragile-X syndrome	14
	Haemophilia A and B	13
	Rett syndrome	4
Mitochondria	Mitochondrial myopathies	2

a risk to the mother's physical or mental health (grounds A, B, E–G).

For further details see the Abortion Law Act 1967, and the Human Fertilisation and Embryology Act (HFEA) 1990.

Genetic testing for non-disease traits

The completion of the mapping of the human genome also opens the door for screening of non-disease traits, for instance sex, hair colour, athletic endurance and IQ. The assumption often made is that certain phenotypes will be eradicated in a drive for genetic 'perfection'. However, such tests also allow for the selection of mutations associated with disease and disability, such as positive selection for deafness, blindness and dwarfism. Such decisions are usually deeply imbedded in culture and trigger huge debate about the rights of the unborn child.

Data protection and ownership of genetic data

Examples of the concerns surrounding the generation, protection and ownership of genetic data can be found in Figure 9.10.

Fig. 9.10 Data protection and ownership of genetic data.

Issue	Examples
The misuse of test results	Insurance provision may be limited if a person is shown by genetic test to have a predisposition to chronic ill health and reduced life expectancy. The potential is for the creation of a genetic underclass.
Rare mutations with the potential for identifying individuals	Disclosure of data pertaining to very rare mutations and conditions has the potential of breaching a patient's right to confidentiality.
Privacy of genetic information	Who should have access and how widely should genetic data be shared, both within academia and industry?
Ownership of genes and chromosomes	Can you patent life itself? Biomedical research is expensive and commercial companies and their shareholders expect a return on their investment. Some have succeeded in doing so, 'selling' genes to large pharmaceutical companies for hundreds of millions of dollars. Documents such as the Universal Declaration on the Human Genome and Human Rights, although having no basis in law, aim to promote the benefits of altruistic conduct in relation to genetic data – a major principle of the HGP was that the sequence results should be placed in the public domain within 24 hours of their being obtained.

SELF-ASSESSMENT

Extended-matching questions (EMQs)

1. Theme: viral genomes

Match the description of the viral genome, or the name of the virus to the correct term used to describe its genome. Each answer can be used once, more than once or not at all.

1. The molecule consists of a 'plus' and a 'minus' strand
2. Hepatitis A and C
3. Viruses that produce a DNA intermediate
4. Measles and mumps
5. The replication of this molecule involves an intermediate mRNA molecule

   A. Positive-sense, single-stranded RNA
   B. Negative-sense, single-stranded RNA
   C. Positive-sense, single-stranded DNA
   D. Negative-sense, single-stranded DNA
   E. Single-stranded RNA
   F. Single-stranded DNA
   G. Double-stranded RNA
   H. Double-stranded DNA
   I. Reverse transcribing RNA
   J. Reverse transcribing DNA

2. Theme: antimicrobial agents

Match the name of the antimicrobial agent to the correct mechanism of action. Each answer can be used once, more than once or not at all.

1. Glycopeptides
2. Rifampicin
3. Erythromycin
4. Quinolones
5. Sulphonamides

   A. Inhibitor of RNA polymerase activity
   B. Inhibitor of cell-membrane function
   C. Inhibitor of cell-wall synthesis
   D. Inhibitor of protein synthesis
   E. Inhibitor of RNA gyrase
   F. Inhibitor of DNA gyrase
   G. Anti-metabolites
   H. Inhibitor of DNA polymerase activity

3. Theme: prokaryotic organelles

Match the prokaryotic organelle to the correct description. Each answer can be used once, more than once or not at all.

1. Ribosome
2. Plasma membrane
3. Nucleiod
4. Cell wall
5. Cytoskeleton

   A. Is membrane bound
   B. In most prokaryotes, contains hopanoids
   C. Major components are FtsZ and MreB
   D. Major components are tubulin and actin
   E. Usually a peptidoglycan structure
   F. Smaller size to its eukaryotic counterpart
   G. Is free in the cytoplasm

4. Theme: eukaryotic organelles

For each of the below, select the eukaryotic organelle being described. Each answer can be used once, more than once or not at all.

1. An organelle required for the degradation of fatty acids
2. This part of the cytoskeleton comprises a tubulin
3. A major function of this organelle is oxidative phosphorylation
4. This organelle is involved in the manufacture of steroid hormones
5. This organelle is the site of rRNA biosynthesis

   A. Rough (granular) endoplasmic reticulum
   B. Mitochondria
   C. Peroxisomes
   D. Nucleus
   E. Microfilaments
   F. Smooth (agranular) endoplasmic reticulum
   G. Golgi apparatus
   H. Nucleoli
   I. Microtubules
   J. Lysosomes
   K. Centrioles
   L. Ribosomes

5. Theme: cell specialization

For each of the below, select the cell specialization being described. Each answer can be used once, more than once or not at all.

1. Muscle tissue
2. Nervous tissue
3. Connective tissue – fibroblasts
4. Epithelial tissue
5. Connective tissue – adipocytes

   A. Neurons are specialized to receive and transmit information

   B. These cells are bound together by tight junctions and form an impermeable barrier

   C. In wound healing these cells bring about scar tissue shrinkage

   D. The rearrangement of external bonds between fibrillar proteins causes contraction

   E. These cells are found in clumps in loose connective tissue

   F. Neurons are specialized to generate information

   G. The rearrangement of internal bonds between fibrillar proteins causes contraction

   H. In wound healing these cells bring about scar tissue degradation

   I. These cells are found in well-arranged masses in fixed connective tissue

   J. These cells are bound together by tight junctions and form a selectively permeable barrier

6. Theme: receptors and signalling

For each of the below, select the type of receptor and signalling modalities being described. Each answer can be used once, more than once or not at all.

1. This molecule is an example of an enzyme-linked receptor
2. This receptor directly affects the activity of a cell by opening the ion channel
3. This receptor activates the inositol lipid pathway
4. Activated intracellular receptors bind to this
5. In this method of cell signalling, the secretory cell is also the target cell

   A. Metabotropic G protein-coupled receptor

   B. G_i

   C. Ionotropic receptor

   D. Apocrine

   E. Hormone sensitive lipase

   F. G_l

   G. Endocrine

   H. Tyrosine-kinase associated receptor

   I. G_s

   J. Hormone response elements

   K. Autocrine

   L. Receptor guanylyl cyclases

   M. G_q

7. Theme: enzyme linked receptors

For each of the below, select the enzyme linked receptor being described. Each answer can be used once, more than once or not at all.

1. These receptors have no catalytic activity and the binding of ligands such as α-interferon induces a noncovalent association of two separate cytokine receptor subunits
2. Activation of these receptors causes the phosphorylation of serine/threonine residues, eventually activating an intracellular second messenger involving the TGF-β superfamily
3. The mechanism of activation of this receptor class begins with a two-stage reaction yielding Pi from the auto-phosphorylation of a residue. This is also known as two-component activation and is used in bacterial chemotaxis
4. After this receptor is activated, tyrosine residues on cytosolic signalling proteins are dephosphorylated. This process is implicated in the activation of leucocyte common antigen, CD45, when cross-linked by extracellular antibodies
5. Atrial natriuretic peptide counters blood pressure increases from the renin–angiotensin system via this class of enzyme-linked receptor. This enzyme catalyzes the production of cGMP, eventually resulting in the phosphorylation of residues in target proteins

   A. Histidine-kinase-associated receptors

   B. Receptor guanylyl cyclases

   C. Receptor serine/threonine kinases

   D. Receptor tyrosine kinases

   E. Tyrosine-kinases associated receptors

   F. Receptor-like tyrosine phosphatases

8. Theme: cellular proteins

For each of the below, select the cellular protein being described. Each answer can be used once, more than once or not at all.

1. A glycosaminoglycan
2. These proteins are only found in the nucleus
3. A type of intermediate filament
4. Lysosomal proteases
5. Expressed exclusively on leucocytes

A. Globular actin
B. β-1 integrin
C. β-tubulin
D. Keratin
E. Lamins
F. Ankyrin
G. Kinesin
H. Cathepsins
I. β-galactosidase
J. Cadherins
K. β-2 integrin
L. Chondroitin sulphate

9. Theme: extracellular matrix

For each of the below, select the component of the extracellular matric being described. Each answer can be used once, more than once or not at all.

1. This fibrous protein is resistant to stretching and has great tensile strength
2. A highly glycosylated, hydrophobic protein, rich in the non-hydroxylated forms of proline and glycine
3. This is an adhesive glycoprotein dimer, involving the promotion of cell proliferation and migration
4. This protein can promote or inhibit adhesion and migration and is exclusively produced by embryonic tissue and glial cells
5. These structures are very hydrophilic, with an extended coil structure that takes up extensive space

   A. Glycoproteins
   B. Collagen
   C. Integrins
   D. Fibronectin
   E. Proteoglycans
   F. Selectins
   G. Cadherins
   H. Elastin
   I. Immunoglobulins
   J. Laminin
   K. Tenascin

10. Theme: amino acid side-groups

For each of the below, match the amino acid to its corresponding side-group. Each answer can be used once, more than once or not at all.

1. This amino acid side-group has a closed-ring structure
2. This side-group is a weak acid, with a pH of 4 and a negative charge
3. This side-group contains sulphur
4. This weakly basic side-group has a pH of 12 and a positive charge
5. This polar side-group is hydroxylic

   A. Aspartic acid
   B. Glycine
   C. Methionine
   D. Alanine
   E. Proline
   F. Arginine
   G. Serine
   H. Valine
   I. Tyrosine
   J. Histidine
   K. Isoleucine
   L. Tryptophan

11. Theme: chromatin: DNA, RNA and their associated nucleoproteins

For each of the below, select the correct name for the type of nucleic acid or associated nucleoprotein being described. Each answer can be used once, more than once or not at all.

1. It is thought that the loops of these structures form transcriptional units
2. On electron micrographs these structures appear as 'beads' on a string of DNA
3. These DNA structures are tandem repeats of a hexameric sequence at the chromosomes' end
4. The level of DNA packing associated with this structure is mediated by histone HI
5. Chromatin is maximally condensed in these structures and can be visualized under light microscope

   A. Centromeres
   B. Nucleosome
   C. Introns
   D. Solenoid
   E. Heteronuclear RNA
   F. Promoters
   G. Giant supercoil
   H. Chromosomes
   I. Telomeres

12. Theme: nuclear genes

For each of the below, select the nuclear genes being described. Each answer can be used once, more than once or not at all.

1. Due to this phenomenon, some functionally related proteins are clustered together on the same chromosome
2. These structures are rare in prokaryotic genes and uncommon in lower eukaryotes
3. These structures are thought to have arisen from incomplete duplication events and functionally redundant gene silencing
4. This feature of nuclear genes is important for terminating transcription and exporting mRNA from the nucleus
5. 'Consensus' sequences are vital for the function of this feature in both eukaryotes and prokaryotes

   A. Active chromatin
   B. Intron shuffling
   C. 3′ sequences
   D. Gene duplication
   E. Exons
   F. Replication forks
   G. Splicing
   H. Introns
   I. HLA complex
   J. Telomerase
   K. Promotors
   L. Methylation
   M. Pseudogenes

13. Theme: detection of target molecules

For each of the below, select the detection tool being described. Each answer can be used once, more than once or not at all.

1. Simultaneous analysis of expression of many genes
2. Quantitation of a protein in many samples
3. Detection of target RNA by incubating probes with a nylon membrane
4. Detection of target DNA by the use of thermal cycling and oligonucleotide primers
5. Identification of specific chromosomes using controlled protein digestion and Giemsa staining

   A. ELISA
   B. G-banding
   C. FISH
   D. Microarray analysis
   E. Southern blotting
   F. Northern blotting
   G. Western blotting
   H. MPLA
   I. CGH
   J. PCR

14. Theme: cloning terminology

For each of the following descriptions, select the most appropriate term. Each answer can be used once, more than once or not at all.

1. A set of overlapping clones
2. Set of clones containing DNA produced by reverse transcription
3. Cloning vector for fragments of up to 350 kb in size
4. Used to 'cut' DNA
5. Virus vector which infects bacteria

   A. Yeast artificial chromosome
   B. Phage
   C. Genomic DNA library
   D. Contig
   E. Bacterial artificial chromosome
   F. Plasmid
   G. Clone map
   H. cDNA library
   I. DNA ligase
   J. Restriction endonuclease

15. Theme: features of chromosomal disorders

For each set of signs, select the most likely diagnosis. Each answer can be used once, more than once or not at all.

1. Low set ears and rockerbottom foot
2. Short stature and webbed neck
3. Occasionally aggressive in childhood
4. Single palmer (simian) crease
5. Tall male with small, soft testes

   A. 47, XYY
   B. Down syndrome
   C. Patau syndrome
   D. Edwards syndrome
   E. Trisomy X syndrome
   F. Klinefelter syndrome
   G. de la Chapelle syndrome
   H. Turner syndrome

16. Theme: inheritance of genetic diseases

For each of the below, select the most appropriate condition. Each answer can be used once, more than once or not at all.

1. X-linked recessive
2. Autosomal recessive cancer syndrome
3. Autosomal recessive, part of Guthrie screening test
4. Autosomal dominant
5. Multifactorial

   A. Duchenne muscular dystrophy
   B. Multiple endocrine neoplasia
   C. Ataxia talangiectasia
   D. Rheumatoid arthritis
   E. Vitamin D resistant rickets
   F. Phenylketonuria
   G. Familial hypercholesterolaemia

17. Theme: prenatal diagnosis

For each of the statements below, chose the most appropriate prenatal diagnostic technique. Each answer can be used once, more than once or not at all.

1. Liquor is removed via a transabdominal needle
2. Associated with the highest rate of miscarriage
3. Routine procedure performed first at 16–18 weeks gestation in all pregnancies
4. Maternal serum screening test for trisomy 21
5. All testing centres for this procedure must be licensed by HFEA

A. Amniocentesis
B. Pre-implantation genetic diagnosis
C. Chorionic villus sampling
D. Cordocentesis
E. Fetoscopy
F. Non-stress test
G. Ultrasound
H. Quadruple test
I. Transcervical trophoblastic cell retrieval

18. Theme: presentations of genetic disease

For each of the presenting symptoms below, choose the most likely diagnosis. Each answer can be used once, more than once or not at all.

1. Cleft lip and palate
2. Skin blisters
3. Meconium ileus
4. Cirrhosis
5. Polycystic kidney disease

   A. Fanconi anaemia
   B. Epidermolysis bullosa
   C. Noonan sysndome
   D. DiGeorge syndrome
   E. Edwards syndrome
   F. Wilson disease
   G. Tuberous sclerosis
   H. Cystic fibrosis
   I. Gilbert disease
   J. Pierre Robin syndrome

Single best answer questions (SBAs)

1. Theme: transfer of genetic material

Select the statement that best describes the transfer of genetic material.

A. Transformation is referred to as the bacterial equivalent of sexual reproduction
B. F plasmids are usually integrated into the bacterial genome
C. Conjugation is referred to as the bacterial equivalent of sexual reproduction
D. Incompatible DNA released by bacteria during transformation is degraded by endonucleases
E. F' plasmids are usually integrated into the bacterial genome

2. Theme: DNA polymerases

Select the statement that best describes DNA polymerases.

A. Will not pause if an incorrect base in inserted
B. Incorporates each nucleotide by coupling it onto the free 3'-OH group
C. Always requires a primer on which to complete extension
D. Removes PP_i from the free phosphate group
E. Incorporates each nucleotide by coupling it onto the free 5' phosphate group

3. Theme: pathogenesis of viral infection

Select the statement that best describes pathogenesis of viral infection.

A. Lytic viral infection leads to cervical and anogenital cancer
B. Persistent viral infection causes genital warts
C. Persistent viral infection completes the replication cycle
D. Lysogenic viral infection cannot result in latency
E. Persistent viral infection can result in transformation

4. Theme: antiviral chemotherapy

Select the statement that best describes antiviral chemotherapy.

A. Aciclovir prevents viral attachment to the host cell
B. Anti-viral agents are least effective when the virus is replicating
C. The virus penetrates the host within an endocytic vesicle after viral nucleic-acid synthesis
D. Inhibiting virus particle maturation prevents the viral genome from being released and taking over host translational machinery
E. Reverse transcriptase inhibitors are important in the modern treatment of HIV

5. Theme: the DNA replication fork

Select the statement that best describes the DNA replication fork.

A. The leading strand can be synthesized in a continuous process from a single RNA primer
B. Three replication complexes form at each origin of replication and proceed in opposite directions
C. Replication is initiated by Okazaki fragments giving rise to two replication forks
D. Nucleotides can only be added to the 5' end of the replication fork
E. DNA ligase adds nucleotides to the 3' end of the replication fork

6. Theme: viruses

Select the statement that best describes viruses.

A. Viruses carry their own metabolic intermediates within the virion
B. Viral nuclear material is enclosed in a phospholipid coat called a capsid
C. The replication of RNA viral genomes is error prone
D. There are many variations of the major stages of replication between viruses
E. Viruses commonly contain both DNA and RNA

7. Theme: HIV

Select the statement that best describes HIV.

A. HIV cannot be transmitted by breastfeeding
B. Dendritic cells are affected by HIV
C. There is only one type of human immunodeficiency virus
D. HIV specifically targets CD4-negative cells
E. HIV is a double-stranded DNA virus

8. Theme: mitochondria

Select the statement that most correctly describes mitochondria.

A. Mitochondria possess their own flagellae
B. Mitochondria have their own double stranded DNA
C. Mitochondria are implicated in the pathogenesis of Alzheimer disease
D. Mitochondria are fully autonomous
E. Mitochondria have nuclear pores small enough to prevent the passage of RNA molecules

9. Theme: the nucleus

Select the statement that most correctly describes the nucleus of a eukaryotic cell.

A. Is bound by a single membrane which is, at points, continuous with the endoplasmic reticulum
B. The main role of nucleoli is the assembly of Golgi bodies
C. RNA passively passes from the nucleus to the cytoplasm
D. Nucleoli are dense staining and one or more of these areas can be present
E. Heterochromatin is light staining and euchromatin is dense staining

10. Theme: the cytoskeleton

Select the statement that most correctly describes the eukaryotic cytoskeleton.

A. Dynein side arms hydrolyse ATP to act as the site of spindle assembly in cell division
B. Flagella are very long microvilli with a '9+2' structure
C. Cilia are used to aid movement of a substance within some cells
D. Microvilli are non-motile extensions of plasma membrane
E. Pseudopodia are extensions of centrioles commonly seen in phagocytes

11. Theme: endoplasmic reticulum

Select the statement that most correctly describes the endoplasmic.

A. Is derived from the trans face of the Golgi apparatus
B. Proteins made within rough endoplasmic reticulum are separate from the proteins packed in the Golgi apparatus
C. Cells that make large amounts of secretory protein have large amounts of smooth endoplasmic reticulum
D. Smooth endoplasmic reticulum has ribosomes attached to its cisternae
E. Liver cells have highly developed smooth endoplasmic reticulum

12. Theme: comparing prokaryotic cells with eukaryotic cells

Select the statement that most correctly describes the differences between prokaryotic cells and eukaryotic.

A. Both prokaryotic and eukaryotic cells are capable of binary fission
B. Specialization is common in both prokaryotic and eukaryotic cells
C. A 50-μm cell is likely to be eukaryotic
D. Prokaryotic cells can exist as independent unicellular organisms or part of multicellular organisms
E. The DNA in eukaryotic cells is linear and free

13. Theme: eukaryotic cell differentiation

Select the statement that most correctly describes eukaryotic cell differentiation.

A. Cell memory is the ability of a zygote to differentiate into all the cell types of an adult organism
B. Differentiation is driven by transcription restriction from sections of the genome
C. Totipotency describes a differentiated cell's ability to retain functional characteristics in a new environment
D. There is no loss of genetic material from somatic cells during development
E. Imprinting is important in transmitting epigenetic changes to daughter cells

14. Theme: vesicular bodies – organelles

Select the statement that most correctly describes membranous organelles comprised of vesicular bodies.

A. Secretory vesicles are derived from the cis face of the Golgi apparatus
B. Macrophages have lysosomes containing hydrolases that activate at a basic pH
C. Peroxisomes are larger than lysosomes
D. The production of insulin is an example of regulated secretory vesicle action
E. Lysosomes are required for the breakdown of toxic compounds

15. Theme: clinical implications of eukaryotic organelles

Select the statement that best describes the role of eukaryotic organelles in clinical medicine.

A. Kartagener syndrome is due to ciliary dyskinesia
B. Enzyme induction by some drugs can lead to the reduction of SER

C. Zellweger syndrome is an autosomal dominant disorder
D. Impaired ciliary function in Kartagener syndrome leads to increased mucous clearance in the lungs
E. The nature of Zellweger syndrome suggests that ribosomes are required for normal CNS function

16. Theme: membrane proteins

Select the statement that most correctly describes membrane proteins.

A. An integral membrane protein can only pass through the cell membrane once
B. Peripheral membrane proteins are associated with both the cytoplasmic and extracellular leaf of the lipid bilayer
C. The cytosolic domains of integral membrane proteins are hydrophobic amino acid rich
D. Peripheral membrane proteins can be attached to the cell membrane electrostatically
E. The α-helical loops of membrane proteins act as markers of self-recognition

17. Theme: properties of biological membranes

Select the statement that most correctly describes biological membranes.

A. The enzyme 'flipase' is required for phospholipid 'flip-flopping'
B. At the transition temperature the cell membrane transforms from a fluid to a structured state
C. Water is not lipid soluble but is membrane permeable due to its charge
D. Protein directionality is unimportant for function
E. Proteins can move in an axial rotational manner, perpendicular to the membrane plane

18. Theme: receptors and drugs

Select the statement that most correctly describes the relationship between drugs and receptors.

A. Antagonists are molecules that bind receptors and active them
B. Agonists can be physiological agents
C. The desired action of a drug commonly arises from binding from several receptor types
D. Reversible antagonists compete with the ligand for the receptor
E. Increasing the concentration of agonist can negate the effect of irreversible antagonist

19. Theme: signal transduction pathways

Select the statement that most correctly describes a concept of transmembrane signalling.

A. All molecules transfer their signal by binding cell surface receptors
B. Different cell types may respond differently to the same signal at the translation level
C. The amplification of signal transduction pathways results in malignancy
D. The variety of responses to second messenger systems is the same between cell types
E. When activated, intracellular receptors migrate and bind to nucleic DNA

20. Theme: membrane potential

Select the statement that most correctly describes membrane potential.

A. The Gibbs-Doonan equilibrium reflects the living cells' sensitivity to osmotic gradients
B. A depolarized nerve cell is more likely to fire
C. E_m is proportional to the concentration of Na^+ on each side of the membrane
D. Hyperpolarization means the E_m becomes less negative
E. The Nernst equation describes the equilibrium between solutions separated by an impermeable membrane

21. Theme: transport mechanisms

Select the statement that most correctly describes cell membrane transport mechanisms.

A. Defects in the glucose transporter cause muscular dysfunction
B. Carrier proteins undergo conformational change before transport
C. Active transporters are carrier proteins that can be linked to an ionic gradient
D. All glucose is transported by facilitated diffusion through uniports
E. Polycystic kidney disease is due to malformation of cellular transport mechanisms

22. Theme: intracellular receptors

Select the statement that most correctly describes intracellular (steroid) receptors.

A. Steroid receptors have both a ligand-binding and a DNA-binding domain
B. Class II steroid receptors are held in complexes with other proteins
C. Tyrosine-kinase-associated receptors are intracellular (steroid) receptors
D. Class I steroid receptors are located in the cell nucleus
E. These are single-pass trans-membrane proteins

23. Theme: Na$^+$/K$^+$ ATPase pump

Select the statement that most correctly describes the sodium/potassium pump.

A. Binding of Na$^+$ causes dephosphorylation and the subunit returns to its original state
B. The β-subunit is the catalytic unit with a K$^+$ binding site in its intracellular surface
C. This pump maintains the high level of intracellular Na$^+$ level
D. This active transporter consists of a glycosylated β-subunit
E. The sodium gradient is used to drive glucose out of kidney cells

24. Theme: distribution of ions across the cell membrane

Select the statement that most correctly describes the distribution of ions across the cell membrane.

A. The concentration of chloride ions inside the cell is less than that outside the cell
B. Energetically unfavourable concentration gradients are maintained by ion channels
C. There is an equal concentration of amino acids both within and outside the cell
D. Polar molecules rely on non-specific proteins for transport across the cell membrane
E. The electrochemical gradient refers to the thermodynamically unfavourable movement of ions

25. Theme: components of the biological membrane

Select the statement that most correctly describes a component of the biological membrane.

A. The three major phospholipids differ with respect to their number of fatty acids
B. Cholesterol consists of four hydrophilic rings and a hydrophobic hydroxyl group
C. At low temperatures, cholesterol is responsible for an increase in fluidity of the cell membrane
D. The intracellular domains of integral proteins are rich in polar amino acid residues
E. Phospholipids are arranged symmetrically, with their components equally distributed between the inner and outer monolayers

26. Theme: Gibbs–Donnan equilibrium

Select the statement that most correctly describes the Gibbs–Donnan equilibrium.

A. The Gibbs–Donnan equilibrium resembles the reality of living cells as they contain impermeant ions such as K$^+$ and Cl$^-$

B. The Gibbs–Donnan equilibrium does not account for the influence of osmotic gradients on living cells, the effect of which is negated by Cl$^-$/K$^+$ exchange pumps
C. The Gibbs–Donnan equilibrium can be used to calculate the direction of flow when the reaction is not in equilibrium in terms of a given potential difference
D. The Gibbs–Donnan equilibrium resembles the reality of living cells as they are not entirely impermeable to positively charged ions
E. The Gibbs–Donnan equilibrium can describe the electrochemical equilibrium that develops when two solutions are separated by a membrane impermeable to both of the ionic species present

27. Theme: disorders of the cell membrane

Select the statement that most correctly describes a disorder of the cell membrane.

A. In cystic fibrosis there is an accumulation of chloride inside the cell and of sodium outside the cell
B. Anti-diuretic hormone leads to a relative increase in serum solute load
C. In cystic fibrosis, the genetic mutations affect a cAMP-regulated gated ion channel
D. Neimann–Pick disease can result from a lack of lysosomal cholesterol
E. Hypokalaemia will tend to partially depolarize excitable cells, causing arrhythmias when affecting cardiac cells

28. Theme: the cytoskeleton

Select the statement that most correctly describes the cytoskeleton.

A. Intermediate filaments have complementary binding sites to enable binding to other monomers
B. Actin and tubulin subunits are symmetrical and wind together
C. Generally, intermediate filaments are associated with cell movement
D. Intermediate filaments have a globular domain at each end
E. Generally, microfilaments are involved with mechanical strength

29. Theme: microfilaments

Select the statement that most correctly describes cellular microfilaments.

A. Myosin proteins move groups of similarly orientated actin filaments past each other
B. Actin facilitates fibroblast movement via rearrangement of its filaments

C. α-actins predominate in non-muscle cells
D. Have a diameter of around 8–11 nm
E. The polymerization of globular G actin subunits cannot be influenced by extracellular signals

30. Theme: protein composition of intermediate filaments

Select the statement that most correctly describes the protein composition of intermediate filaments.

A. Keratins can be found in epithelia lining internal body cavities
B. Neurofilaments are found in glial cells surrounding neurons
C. Lamins are exclusive to the nuclear lamina
D. Cells usually contain more than one type of intermediate filament
E. Desmin is expressed predominantly in mesenchymal cells

31. Theme: microtubules

Select the statement that most correctly describes microtubules in the working cell.

A. Microtubules are lacking in neurons as intermediate filaments direct axon elongation
B. Microtubule-associated proteins have specific interactions with kinesin and dynein
C. Microtubules hollow tubules composed of vimentin dimer polymers
D. Microtubules-associated proteins can function as ATP-independent molecular motors
E. Microtubules can produce movement through ATP-powered dynein arm linkage

32. Theme: lysosomes in the working cell

Select the statement that most correctly describes the role of lysosomes in the working cell.

A. Lysosomes enable heterophagy; recycling products of receptor-mediated endocytosis
B. Erythrocytes contain hundreds of lysosomes, whereas phagocytes have thousands
C. Primary lysosomes are formed from the fusion of a lysosome with a vesicle containing substrate
D. Endocytosis can be both specific and non-sepcific
E. The ATP driven H^+ pump maintains the lysosomal matrix pH at 7.5–8.5

33. Theme: cell-to-cell junctions

Select the statement that most correctly describes cell-to-cell junctions.

A. Adherans junctions occur as streak-like attachments or continuous bands
B. Cell-to-cell contact at communicating junctions is mediated by occluding and claudin
C. Occluding junctions are most abundant in cells under stress, such as cardiac muscle
D. The degree of permeability offered by tight junctions is influenced by extracellular signals
E. Inorganic ions are able to pass through anchoring junctions to permit cell coupling

34. Theme: adhesion molecules

Select the statement that most correctly describes adhesion molecules.

A. Selectins are generally involved in Ca^{2+} interactions
B. Integrins composed the adhesion belt and desmosomes
C. Selectins are only present in blood and endothelial cells
D. Immuniglobulins are associated with actin filaments
E. Cadherin expression is induced by local chemical mediators

35. Theme: extracellular matrix

Select the statement that most correctly describes a component of the extracellular matrix.

A. Fibrillar collagen is capable of stretching whereas non-fibrillar is stretch resistant
B. Laminin binds to secreted signalling molecules and provides hydrated space
C. Proteoglycans have a greater proportion of carbohydrate to glycoproteins
D. Type IV collagen is nonfibrillar and is only found as a sheet in the basal lamina
E. Elastin is a highly glycosylated, hydrophilic protein found in tissues requiring flexibility

36. Theme: the role of the fibroblast

Select the statement that most correctly describes the role of the fibroblast in connective tissue.

A. Fibroblasts secrete fibrous protein components of the ECM in all tissues
B. Fibroblasts are interchangeable with other members of the connective tissue family under appropriate conditions
C. Fibroblasts anchor the cell surface to the basement membrane
D. Fibroblasts disable the configuration of ECM into tendons and other structures
E. Fibroblasts act as sieves to regulate molecular trafficking

37. Theme: essential and non-essential amino acids

Select the statement that most correctly describes essential and/or non-essential amino acids.

A. There are twelve essential amino acids in the human body
B. The semi-essential amino acids are histidine and arginine
C. Non-essential amino acids are required in the diet
D. Valine has the simplest side group; hydrogen
E. The free carboxyl group is considered the start of a polypeptide sequence

38. Theme: properties of amino acids

Select the statement that most correctly describes a property of amino acids.

A. Large, bulky amino acid side chains are responsible for 'steric hindrance'
B. In aqueous solution globular colloidal proteins have polar groups arranged on the inside
C. Non-covalent hydrophobic bonds occur between a hydrophobic and a hydrophilic residue
D. Side groups cannot be modified in post-translational reactions
E. Amino acids with non-polar R groups form amphions at a neutral pH

39. Theme: forces that shape proteins

Select the statement that most correctly describes a force that shapes protein structure.

A. Peptide bonds are shorter in length than hydrogen bonds
B. The peptide bond confers complete rigidity to the polypeptide chain
C. Hydrophobic residues are able to form true bonds through water displacement
D. Together, Van der Waals forces add little energy in the polypeptide tertiary structure
E. Disulphide bridges are favoured intracellularly rather than in exported proteins

40. Theme: structures within proteins

Select the statement that most correctly describes the structures within proteins.

A. Zinc finger proteins influence translation by binding to specific DNA sequences
B. Each amino acid of the α-helix participates in a hydrogen bond
C. Prions that produce disease have structures dominated by β-sheets
D. Three polypeptide chains linked by hydrophobic interactions compose the collagen helix
E. Fibrous proteins are mostly made up of β-pleated sheets

41. Theme: mechanism of enzyme action

Select the statement that most correctly describes the mechanism of enzyme action.

A. In comparison to a catalyzed reaction, a non-catalyzed reaction has less of a change in energy between reactants and products
B. The transition state is the same for catalyzed and non-catalyzed enzymatic reactions
C. The 'induced fit' model proposes that as the product dissociates the enzyme remains in a state of conformational change
D. Reduced enzymatic reaction rates at over $37°C$ is due to a disruption of the enzyme's active site
E. There is a different position of equilibrium when comparing a catalyzed and a non-catalyzed reaction

42. Theme: enzyme kinetics

Select the statement that most correctly describes enzyme kinetics.

A. Allosteric inhibitors bind to a site other than the active site but still can induce conformational change
B. The Michealis–Menton equation is used to calculate the rate of breakdown of the enzyme-substrate complex
C. Denaturing treatments and irreversible inhibitors are good examples of enzyme inhibitors
D. The Michaelis constant (K_m) relates the reaction rate to the substrate concentration
E. The Michaelis–Menton graph is used to estimate V_{max} and K_m

43. Theme: carbohydrates

Select the statement that most correctly describes a type of carbohydrate.

A. Disaccharides are formed by joining two monosaccharides and eliminating hydrogen
B. Polysaccharides are polymers with the general formula $C_n(H_2O)_{n+1}$
C. Oligosaccharides often covalently attach to proteins or membrane lipids
D. Glycogen is a highly branched heteropolysaccharide
E. In humans, most carbohydrates exist in the L stereoisomer form

44. Theme: sugar derivatives

Select the statement that most correctly describes a sugar derivative.

A. Sugar acids are formed by the reduction of a carbonyl group
B. Deoxyribose and fucose are examples of nucleosides
C. Sugar alcohols are formed by the oxidation of the C1 aldehyde group or the hydroxyl on the terminal carbon
D. Deoxy sugars are formed by substituting a hydroxyl group for hydrogen
E. Adenosine, deoxyadenosine and thymidine are all examples of amino sugars

45. Theme: functions of macromolecules

Select the statement that most correctly describes a macromolecule function.

A. The major histocompatibility complex is an example of a protein used for communication
B. Lactose intolerance is the inability of gastrointestinal bacteria to ferment lactose
C. The substitution of bulkier amino acids for glycine is implicated in osteogenesis imperfecta
D. Glycogen is highly soluble in the cell, therefore accumulation of large amounts contributes to cell lysis
E. Myoglobin has eight β-pleated sheet segments with which to bind oxygen in muscle

46. Theme: macromolecules in health and disease

Select the statement that most correctly describes a macromolecule in either health or disease.

A. Haemoglobin A is a tetramer composed of three α-subunits and a β-subunit
B. Enzyme-catalyzed reactions show maximum activity at 35°C
C. A disruption of collagen architecture and quality has been implicated in osteoporosis
D. Alcohol dehydrogenase requires the presence of the coenzyme NAD
E. Following denaturation and renaturation, functional insulin is recoverable

47. Theme: structures of the nucleus

Select the statement that most correctly describes a feature of nuclear structure.

A. At least one nucleoli is always visible in the nucleus of a cell
B. Nuclear pores are electron dense structures arranged around the periplasm
C. Histones are strongly acidic globular proteins
D. A thin chromatic shell underlies the inner nuclear membrane
E. Transport of all molecules across the nuclear pores is ATP-dependent

48. Theme: nucleotides and nucleosides

Select the statement that most correctly describes these types of nucleic acid.

A. Eukaryotic cells cannot synthesize purines and pyrimidines *de novo*
B. A nucleoside is composed of a base, a sugar ad a phosphate molecule
C. Nucleotides can be synthesized from the breakdown of exogenous nucleic acids
D. Cytosine, thymine and uracil are purines
E. Folate derivatives are implicated in purine biosynthesis

49. Theme: RNA molecules

Select the statement that most correctly describes an RNA molecule.

A. mRNA is derived from splicing an initial RNA transcript in prokaryotes and eukaryotes
B. Ribosomal RNA is the primary mRNA transcript produced by eukaryotic cells
C. The majority of tRNA bases undergo post-translational modification
D. mRNA has a 'clover-leaf'-shaped secondary structure due to intromolecular base pairing
E. Heteronuclear RNA carries specific amino acids to the site of protein synthesis

50. Theme: the genetic code

Select the statement that most correctly describes the genetic code.

A. There are 61 possible codons in genetic translation
B. The tRNA anticodons bind a specific base only
C. There is one codon encoding each amino acid
D. There are 61 amino-acid coding codons
E. Mitochondrial DNA is the same as that in the rest of the cell

51. Theme: post-translational protein modification

Select the statement that most correctly describes post-translational protein modification.

A. The signal peptide is important in marking proteins for export from the cell
B. The polypeptide can undergo post-translational modification inside the ER lumen
C. Sulphation regulates enzyme or protein activity
D. Proteins are directed to the correct post-translational location by peptide cleavage
E. In phosphorylation, kinases transfer phosphate groups from the target residue onto ATP

52. Theme: regulation of the cell cycle

Select the statement that most correctly describes regulation of the cell cycle.

A. 'Checkpoint proteins' are effective against DNA damage, not structural problems
B. Cyclins function independent of mRNA expression and protein levels
C. Activated CDKs stimulate the cell cycle progression by phosphorylating cellular proteins
D. CDK activity cannot be influenced by extracellular signalling pathways
E. Growth factors regulate the cell cycle at a local level only

53. Theme: the cell cycle and cancer

Select the statement that most correctly describes the cell cycle in terms of cancer.

A. Cancer is due to either an increase or a decrease in expression of genes associated with the cell cycle
B. Integration of viral DNA into the host genome is a rare cause of cancer
C. Failure of tumour suppressor genes results in the imortalization of DNA damage in the genome
D. Oncogenes are only expressed in malignant cells
E. Cancer can exist before the genomic changes confer a change in the cell's phenotype from normal to malignant

54. Theme: mitosis

Select the statement that most correctly describes the process of mitosis.

A. The end of telophase is marked by the clustering of a complete set of chromosomes at each pole
B. Nuclear envelope vesicles migrate towards the poles during prophase
C. Kinetochores are formed during prophase
D. Chromosomes are maximally condensed at the end of metaphase
E. Anaphase is when the centrioles are pulled to opposite poles by the spindle

55. Theme: DNA mutation

Select the statement that most correctly describes the process of DNA mutation.

A. X-rays are mutagens whereas γ-rays are safe
B. Base analogue chemical mutations cause misreading of the DNA
C. Some antibiotics are chemical modifying mutagens
D. Viral genomes can disrupt coding regions but have no impact on expression of existing genes
E. The products of normal metabolism are protective against DNA mutation

56. Theme: meiosis

Select the statement that most correctly describes meiosis.

A. During the zygotene stage of homologous recombination, bivalents are formed

B. During anaphase I the chromatids separate and migrate to opposite poles of the spindle
C. Telophase I results in the formation of two genetically identical haploid cells
D. Diplotene is the stage of homologous recombination in which the spindle forms
E. During the second division, the chromatids separate in metaphase II

57. Theme: DNA-RNA hybridization

Which of the following molecular tools involves DNA-RNA hybridization?

A. ELISA
B. FISH
C. Northern blotting
D. CGH
E. Mass spectroscopy

58. Theme: microsatellites

Which of the following best describes a microsatellite?

A. A region of repetitive DNA sequences at the end of a chromosome
B. A nucleotide sequence within a gene
C. Nucleic acid that determines the amino acid sequence in proteins
D. A type of DNA polymorphism
E. The region of DNA where two identical sister chromatids come closest in contact

59. Theme: comparative genetic hybridization

Which of these best describes comparative genetic hybridization?

A. This technique highlights areas of DNA with high sequence similarity
B. This technique stains condensed chromosomes
C. This technique is the most appropriate to identify small duplications of chromosomal DNA
D. This technique employs copy number detection and methylation profiling
E. This technique uses labelled sequence-specific DNA probes consisting of oligonucleotides

60. Theme: RNA compliment

Which of the following is most concerned with the full RNA compliment?

A. Genomics
B. Transcriptomics
C. Proteomics
D. Metabolomics
E. Cytomics

61. Theme: recombination fractions

Which of the following is the most likely recombination fraction for two unlinked loci?

A. 0
B. 0.25
C. 0.5
D. 0.75
E. 1

62. Theme: therapeutic monoclonal antibodies

Which of the following is a therapeutic monoclonal antibody?

A. Infliximab
B. Spironilactone
C. Amlodipine
D. Cyclizine
E. Ropinirole

63. Theme: the human genome

The human genome is estimated to consist of approximately how many genes?

A. 15 000–30 000
B. 30 000–45 000
C. 45 000–60 000
D. 60 000–75 000
E. 75 000–90 000

64. Theme: producing monoclonal antibodies

Which of the following is used to produce monclonal antibodies?

A. Bacteria
B. Yeast
C. Hybridomas
D. Human cell line
E. Goat

65. Theme: genetic technique

Which of the following techniques uses pedigree data to determine the map distance between two markers on a chromosome?

A. Polymorphic marker analysis
B. Genetic linkage analysis
C. Exome sequencing
D. Candidate gene approach
E. Functional cloning

66. Theme: alternative splicing

Which of the following best describes alternative splicing?

A. A type of post-translational modification
B. The use of alternate separating enzymes during DNA replication
C. The cross-over of material between chromosomes
D. Production of multiple mRNA transcripts from a single DNA sequence
E. The random assortment of chromosomes at anaphase

67. Theme: mode of inheritance

Using the diagram below, select the mode by which the genetic trait is inherited.

A. Autosomal dominant
B. Autosomal recessive
C. X-linked dominant
D. X-linked recessive
E. Y-linked

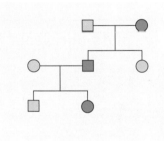

68. Theme: genetic mutation

In sickle-cell disease, a single substitution of an 'A' by 'T' on the β-globin gene causes the resultant protein to have altered properties, but to be of normal size. Select the term that most accurately describes this type of mutation.

A. Expansion mutation
B. Silent mutation
C. Missense mutation
D. Nonsense mutation
E. Frameshift mutation

69. Theme: horizontal inheritance

Select the mode of transmission described by the term 'horizontal inheritance'.

A. Autosomal dominant
B. Autosomal recessive
C. X-linked dominant
D. X-linked recessive
E. Y-linked

70. Theme: autosomal recessive inheritance

A carrier of an autosomal recessive disorder has an affected partner. Select the risk to each pregnancy of producing an effected child.

A. 1:4
B. 1:3
C. 1:2
D. 3:4
E. 1:1

71. Theme: genetic terminology

Select the terminology that may be described as 'differential expression of genetic material depending on which parent it was inherited from'.

A. Anticipation
B. Imprinting
C. Uniparental disomy
D. Mitochondrial inheritance
E. Mosaicism

72. Theme: proto-oncogenes

From the following, select the proto-oncogene.

A. Rb
B. myc
C. p53
D. APC
E. NF-1

73. Theme: Gorlin syndrome

Select the tumour most characteristic of Gorlin syndrome.

A. Small cell carcinoma
B. Endometrial carcinoma
C. Hepatocellular carcinoma
D. Thyroid carcinoma
E. Basal cell carcinoma

74. Theme: isochromosome

Select the statement that best describes an isochromosome.

A. The transfer of one segment of a chromosome to another
B. When a chromosome breaks in two places and the resulting DNA fragment is reversed and re-inserted into it
C. The deletion of one arm of a chromosome, with duplication of the other
D. Abnormal copying of a part of a chromosome
E. A chromosome whose arms have fused together

75. Theme: penetrance

Select the statement that best describes penetrance.

A. The proportion of individuals with a given genotype, which show the associated phenotype
B. The number of new cases of a given disease during a given period in a specified population
C. The probability that two individuals will both have a certain characteristic
D. The total number of cases of a given disease in a specified population at a designated time
E. The degree to which a characteristic is determined by our genes

76. Theme: penetrance

Select the technology that relies on homologous recombination of DNA by endogenous enzymes.

A. Antisense oligonucleotides
B. RNAi
C. Zinc finger nucleases
D. Gene therapy
E. Exon skipping

77. Theme: genetic pedigree

Select the symbol for an affected male, when drawing a family tree (pedigree).

A. Filled-in circle
B. Half-filled circle
C. Empty circle
D. Filled-in square
E. Empty square

78. Theme: autosomal recessive inheritance

Select the approximate heterozygote frequency for an AR disorder with a frequency of 1 in 2500 live births.

A. 1/25
B. 1/50
C. 1/75
D. 1/100
E. 1/125

79. Theme: the Guthrie test

Select the condition screened for in the newborn blood spot (Guthrie) test.

A. β-thalassaemia
B. Phenylketonurea
C. Tay–Sachs disease
D. Down syndrome
E. Charcot–Marie–Tooth disease

80. Theme: X-linked recessive inheritance

Shirley's mother is a known carrier of an X-linked recessive condition. Shirley has had 3 unaffected sons. Select the risk that Shirley herself is a carrier.

A. 8:1
B. 3:1
C. 1:1
D. 1:3
E. 1:8

81. Theme: termination of pregnancy

Select the most common statutory grounds on which termination of pregnancy is performed in the UK.

A. A and B
B. C and D
C. E and F
D. G and H
E. I and J

82. Theme: justice

Select the statement that best describes the principle of justice.

A. Seeking to do good
B. Seeking to not do harm
C. Seeking equality of access
D. Seeking to respect autonomy
E. Seeking to maintain confidentiality

83. Theme: antenatal testing

Select the maternal serum test that is most likely to yield a high value in trisomy 21, but low value in trisomy 13 and 18.

A. α-fetoprotein (AFP)
B. Human gonadotrophin (HCG)

C. Unconjugated oestradiol (uE3)
D. Dimeric inhibin-A (DIA)
E. Human growth hormone (hGH)

84. Theme: the founder effect

Select the statement that best describes the founder effect.

A. A new disease causing mutation arises *de novo*
B. A gene defect is passed between organisms
C. Disease phenotype is caused by a transcription defect
D. Over-representation of gene defect in a small community
E. The first documented genetic mutation to cause a given phenotype

85. Theme: β-thalassaemia

Select the ethnic group at highest risk of β-thalassaemia.

A. Caucasian
B. Mediterranean
C. South American
D. South Asian
E. Afro-Caribbean

86. Theme: pre-implantation genetic diagnosis

From the following, select the autosomal dominant condition currently licensed for pre-implantation genetic diagnosis (PGD) in the UK.

A. Familial hypercholesterolaemia
B. Ehlers–Danlos syndrome (type I)
C. Hereditary spherocytosis
D. Hereditary haemorrhagic telangiectasia
E. Huntington disease

EMQ answers

1. 1H, 2A, 3I, 4B, 5 J

2. 1C, 2A, 3D, 4F, 5G

3. 1F, 2B, 3G, 4E, 5C

4. 1C, 2I, 3B, 4F, 5L

5. 1G, 2A, 3C, 4J, 5E

6. 1H, 2C, 3M, 4J, 5K

7. 1E, 2C, 3A, 4F, 5B

8. 1L, 2E, 3D, 4H, 5K

9. 1B, 2H, 3D, 4K, 5E

10. 1E, 2A, 3C, 4F, 5G

11. 1G, 2B, 3I, 4D, 5H

12. 1D, 2H, 3M, 4C, 5K

13. 1D, 2A, 3F, 4J, 5B

14. 1D, 2G, 3A, 4I, 5B

15. 1D, 2H, 3A, 4B, 5F

16. 1A, 2C, 3F, 4G, 5D

17. 1A, 2C, 3G, 4H, 5B

18. 1J, 2B, 3H, 4F, 5G

1.	C		44.	D
2.	B		45.	A
3.	B		46.	D
4.	E		47.	D
5.	A		48.	E
6.	C		49.	C
7.	B		50.	D
8.	C		51.	B
9.	D		52.	C
10.	D		53.	A
11.	E		54.	D
12.	C		55.	B
13.	B		56.	A
14.	D		57.	C
15.	A		58.	D
16.	D		59.	C
17.	A		60.	B
18.	D		61.	C
19.	E		62.	A
20.	B		63.	A
21.	C		64.	E
22.	A		65.	B
23.	D		66.	D
24.	A		67.	C
25.	C		68.	C
26.	E		69.	B
27.	C		70.	C
28.	D		71.	B
29.	B		72.	B
30.	A		73.	E
31.	E		74.	C
32.	D		75.	A
33.	A		76.	C
34.	C		77.	D
35.	D		78.	A
36.	B		79.	B
37.	B		80.	E
38.	E		81.	B
39.	A		82.	C
40.	C		83.	B
41.	D		84.	D
42.	A		85.	B
43.	C		86.	E

Objective structured clinical examination questions (OSCEs)

Station 1

You are the FY2 doctor performing neonatal examinations (baby checks) on the maternity ward. You notice that a one day-old baby boy (Tim) has some of the characteristic features of Down syndrome. Explain to the baby's mother (Mrs Hill) that you think her baby may have Down syndrome, and explain what you would like to do next.

Checklist

- Introduces themselves and gains consent to talk to Mrs Hill.
- Suggests talking in a private place *or* requests a few minutes of her *uninterrupted* time.
- Tells Mrs Hill her son *may* have Down syndrome (*no mark* for 'definitely Down syndrome').
- Asks Mrs Hill what she knows about Down syndrome.
- Explains it is a congenital condition *or* a permanent/lifelong condition.
- Describes some features of Down syndrome (e.g. facial features, developmental delay).
- Offers to write down the name of the condition.
- Offers confirmation of the diagnosis by blood test.
- Offers other related tests (e.g. echocardiogram, hearing test, eye tests).
- Offers Mrs Hill the opportunity to ask questions.
- Offers Mrs Hill the opportunity to return again *or* directs Mrs Hill to more information.
- Thanks Mrs Hill for her time.
- Sensitive and courteous throughout.

Prompting questions from Mrs Hill:

- 'Is this permanent?'
- 'Will my baby be normal?'
- 'How can you know for certain that this is Down syndrome?'

Station 2

You are a doctor working in a GP practice. Mr and Mrs Owen have come to see you as their 2-year-old son James has just been diagnosed with cystic fibrosis (CF). They want to know more about the condition and what causes it.

Checklist

- Introduces themselves.
- Establishes a rapport.
- Asks Mr and Mrs Owen what they already know about CF.
- Explains that CF is a genetic condition.
- Explains that CF has no cure/is permanent/requires lifelong treatment.
- Explains at least one effect of cystic fibrosis (e.g. chest infections, malnutrition).
- Explains at least one treatment modality (e.g. prophylactic antibiotics, nutritional supplements).
- Asks if Mr and Mrs Owen have any questions.
- Offers Mr and Mrs Owen the opportunity to return again *or* directs them to more information.
- Thanks Mrs and Mrs Owen for their time.
- Sensitive and courteous throughout.

Mr and Mrs Owen *will* ask:

- 'We are planning on having another child, what is the risk they will have CF?'

  Answer:

- One-in-four/25%.
- Offer PGD.

Prompting questions from Mr and Mrs Owen:

- 'What caused it?'
- 'Is it permanent/can this be cured?'
- 'How will this be treated?'

Station 3

You are a medical student on attachment in a neurology clinic. Mr Thomas is presenting with an 8-month history of clumsiness and poor concentration. Please take a history from him and suggest a diagnosis.

Checklist

- Introduces themselves.
- Gains consent to speak to Mr Thomas.
- Asks Mr Thomas his age (*52 years old*).
- Determines Mr Thomas' ethnic background.

- Asks about any other symptoms Mr Thomas has noticed, *to reveal a fine resting tremor, irritability and mood changes.*
- Asks specifically about family history.
- Elicits that *Mr Thomas' father and grandmother had similar symptoms before they died.*
- Asks specifically about past medical history and current medications.
- Asks specifically about recreational drug and alcohol use.
- Asks specifically about activities of daily living and social history.
- Performs a review of systems.
- Says would like to examine Mr Thomas.
- Suggests correct diagnosis (*Huntington disease*).
- Asks questions with a logical, systematic approach.
- Sensitive and courteous throughout.

Prompting questions from the examiner:

- 'Is there anything else you would like to ask or do?'
- 'What do you think is the diagnosis?'

Station 4

You are a doctor in a GP surgery. Mrs Babu has come to see you because a child from her husband's first marriage has recently been diagnosed with sickle-cell anaemia. She is planning on having a baby but is concerned. She wants to know what the chances are that her baby will be affected.

Note: Sickle-cell anaemia is an autosomal recessive disorder.

Information from Mrs Babu:

Her mother was affected with sickle-cell anaemia. Neither her, her husband, or his first wife are affected with sickle-cell anaemia. She knows of nobody else in the family who is affected.

Checklist

- Introduces themselves.
- Establishes a rapport.
- Determines Mr and Mrs Babu's ethnic descent.
- Asks Mrs Babu what she knows about sickle-cell anaemia and its cause.
- Explores the possibility of non-paternity.
- Explains that sickle-cell anaemia is a genetic disease.
- Explores the family history.
- Draws correct family tree (see below).
- Determines correct probability of the child being affected with sickle-cell anaemia (*one-in-four, 25%*).
- Offers pre-implantation genetic diagnosis.
- Sensitive and courteous throughout.

Prompting questions from Mrs Babu:

- 'What is the chance of my baby being affected?'
- 'Is there any way to know for certain before I get pregnant?'

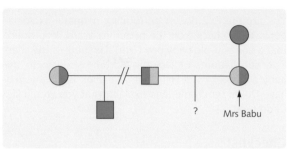

Station 5

You are a medical student in an antenatal unit. Mrs Ladwa is 12 weeks pregnant with her first baby. She is very concerned because she has been offered an ultrasound scan. Please explain to her why the procedure is conducted, and what it will involve.

Checklist

- Introduces themselves.
- Gains consent to speak to Mrs Ladwa.
- Asks what she knows about the procedure already.
- Explores her concerns.
- Explains the procedure is routine.
- Explains that the procedure is painless.
- Explains what the procedure will involve (to mention: supine position, jelly, probe).
- Explains that the procedure will take approximately 5–15 minutes.
- Explains that there is no risk to her baby from the procedure.
- Explains that the scan cannot detect every abnormality.
- Checks that she fully understands the procedure.
- Asks if she still has any concerns or questions.
- Sensitive and courteous throughout.

Prompting questions from Mrs Ladwa:

- 'Does this mean you are worried my baby is abnormal?'
- 'Is there any risk to me or my baby?'
- 'Will this tell me for sure if my baby has an abnormality?'

Allele An allele is one of a series of possible alternative forms of a given gene or DNA sequence at a given locus.

Aneuploidy The condition in which the chromosome number of the cell is not an exact multiple of the haploid number. Monosomies and trisomies are examples of aneuploidy.

Antisense A piece of nucleic acid, typically created in the lab, which has a sequence exactly opposite to an mRNA molecule made by the body. Antisense can bind tightly to its mirror image mRNA, preventing a particular protein from being made.

Autosomal dominant A trait or disease that is produced when only one copy of a polymorphism or mutation is present on an autosome.

Autosomal recessive A trait or disease that is produced when two copies of a polymorphism or mutation are present on an autosome.

Autosomes These are any chromosomes other than the sex chromosomes.

Cell The cell is the basic unit of life. If it is to survive, each cell must maintain an internal environment that supports its essential biochemical reactions, despite changes in the external environment. Therefore, a selectively permeable plasma membrane surrounding a concentrated aqueous solution of chemicals is a feature of all cells.

Clone A member of a group of cells that all carry the same genetic information, and which are all derived from a single ancestor by repeated mitoses.

Compound heterozygote This is an individual with two different mutant alleles at the same locus.

Conjugation The transfer of genetic material between bacteria via direct or bridge-like contact.

Consultand An individual seeking, or referred for, genetic counselling.

Episome DNA which is not part of a chromosome, but is replicated along with the genome.

Exon Region of a gene containing DNA that codes for a protein.

Gamete The reproductive cell formed by meiosis containing half the normal chromosome number.

Genome This is the entire genetic complement of a cell.

Genotype This is the genetic constitution of an individual, and it is also used to refer to the alleles present at one locus.

Germline The cells from which gametes are derived, which contribute to the genetic complement of the offspring.

Heritability The degree to which a characteristic is determined by our genes.

Heterozygote This is an individual or genotype with two different alleles at a given locus on a pair of homologous chromosomes.

Holoenzyme The complete enzyme including all subunits and co-factors.

Homozygote This is an individual or genotype with identical alleles at a given locus on a pair of homologous chromosomes.

Hormone A molecule produced by an endocrine cell, which is released into the bloodstream and acts on specific receptors to elicit its effect.

Host The organism used to propagate a recombinant DNA molecule (usually *E. coli* or *S. cerevisiae*).

Insert The fragment of foreign DNA cloned into a particular vector.

Intron A section of a gene that does not contain any instructions for making a protein.

Karyotype This is the chromosome complement of a cell. In a standard karyotype the chromosomes are conventionally arranged in an order depending upon size. Chromosomes are distinguished individually by their size, centromere position and banding pattern. The normal human karyotype is 46,XY (male) or 46,XX (female).

Library A collection of cloned DNA fragments which, taken together, represent the entire genome of a specific organism. Using traditional techniques they allow the isolation and study of individual genes.

Ligand A molecule, such as a hormone or neurotransmitter, which binds the receptor, and is termed the first messenger.

Locus The position of a gene on a chromosome.

Microarray A large set of cloned nucleic acid molecules or proteins spotted onto a solid matrix, usually a microscope slide, and used to profile gene and protein expression in cells and tissues.

Monosomy A chromosome constitution in which one member of a chromosome pair is missing (e.g. Turner syndrome (45,XO)).

mRNA Messenger RNA which determines the amino acid sequence in proteins

Multifactorial disorder This is a term used to describe disorders in which both environmental and genetic factors are important.

Mutation A mutation is a permanent heritable change in the sequence of DNA.

Neurotransmitter A molecule that is used to transmit nerve impulses across a synapse.

Operon A prokaryotic locus consisting of two or more genes that are transcribed as a unit and are expressed in a coordinated manner.

Organism An organism is a system capable of self-replication and self-repair, which may be unicellular or multicellular. Unicellular organisms consist of a solitary cell able independently to perform all the functions of life. Multicellular organisms contain several different cell types that are specialized to perform specific functions.

Pedigree charts These are used to illustrate inheritance.

Penetrance The proportion of individuals with a specified genotype that show the expected phenotype under defined environmental conditions. The term is usually used in association with dominant disorders.

Phenocopy This is the alteration of the phenotype by environmental factors during development to produce a phenotype that is characteristically produced by a specific gene (e.g. rickets due to a lack of vitamin D would be a phenocopy of vitamin D-resistant rickets).

Phenotype This is the observed biochemical, physiological or morphological characteristics of an individual that are determined by the genotype and the environment in which it is expressed.

Plasmid Autonomously replicating, extrachromosomal circular DNA molecules. Often expoited in the laboratory as a vector for gene cloning.

Pleiotrophy Pleiotrophy is the phenomenon in which a gene is responsible for several distinct and apparently unrelated phenotypic effects, which may concern the organ systems involved and the signs and symptoms that occur.

Ploidy This refers to the number of complete sets of chromosomes in a cell. A haploid cell contains a single set of chromosomes (e.g. gametes); a diploid cell contains two copies of each chromosome (e.g. somatic cells); a polyploid cell contains more than two sets of chromosomes (sometimes this occurs normally in plants and in some animal cells, such as megakaryocytes).

Polygenic inheritance This is a term used to describe the inheritance of traits that are influenced by many genes at different loci.

Polymorphism Polymorphism is the occurrence in a population of two or more alternative genotypes, each at a frequency greater than that which could be maintained by recurrent mutation alone. A locus is arbitrarily considered to be polymorphic if the rarer allele has a frequency of at least 0.01. Any allele rarer than this is a 'rare variant'.

Proband The first person in a pedigree to be clinically identified as having a disease in question.

Probe A probe is a radioactively or fluorescently labelled piece of single-stranded DNA of defined sequence.

Recombinant DNA A molecule of DNA created in vitro that contains elements of more than one original sequence, e.g. vector and insert.

Rolling circle replication A process of nucleic acid replication that can rapidly synthesize multiple copies of circular molecules of DNA such as bacterial chromosomes and plasmids.

Somatic All body cells, except those from which gametes are derived.

Transcription The formation of mRNA from a DNA template.

Transduction The viral transfer of DNA from one bacterium to another.

Transformation The genetic alteration of a cell by DNA uptake.

Translation The formation of a peptide chain from an mRNA template.

Translocation The transfer of one segment of a chromosome to another.

Trisomy The state of having three representatives of a given chromosome instead of the usual pair (e.g. as in trisomy 21 – Down syndrome).

Variable expressivity This occurs when a genetic lesion produces a range of phenotypes: for example, tuberous sclerosis can be asymptomatic with harmless kidney cysts, but in the next generation it may be fatal, owing to the development of brain malformations.

Vector A DNA molecule capable of replicating within a particular host, into which foreign DNA may be inserted.

Wild-type The typical form of an allele as it occurs in nature.

X-linked dominant A trait (or a disease) that is produced when only one copy of a polymorphism or mutation is present on an X chromosome. This means that both males and females can display the trait or disorder, by only having one copy of the gene.

X-linked recessive A trait or disease that is produced when a polymorphism or mutation in a gene on the X chromosome causes the phenotype to be expressed. Remember, females have two X chromosomes, while males have one X and one Y.

Index

Note: Page numbers followed by *b* indicate boxes, *f* indicate figures and *ge* indicate glossary terms.